Hans Peter Good

On the Origin of Natural Constants

Also of Interest

Dynamical Evolution of Galaxies
Xiaolei Zhang, 2018
ISBN 978-3-11-052519-9, e-ISBN 978-3-11-052742-1

Space – Time – Matter
Jochen Brüning, Matthias Staudacher (Eds.), 2018
ISBN 978-3-11-045135-1, e-ISBN 978-3-11-045215-0

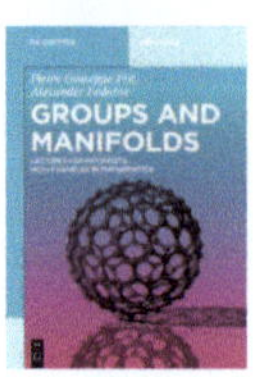

Groups and Manifolds
Lectures for Physicists with Examples in Mathematica
Pietro Fré, Alexander Fedotov, 2017
ISBN 978-3-11-055119-8, e-ISBN 978-3-11-055120-4

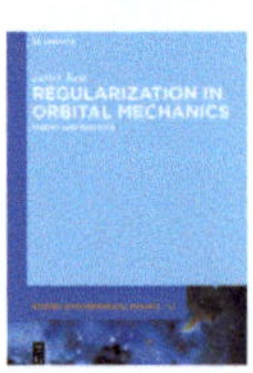

Studies in Mathematical Physics
Volume 42
Regualrization in Orbital Mechanics
Javier Roa, 2017
ISBN 978-3-11-055855-5, e-ISBN 978-3-11-055912-5
ISSN 2194-3532

Hans Peter Good

On the Origin of Natural Constants

Axiomatic Ideas with References to the Measurable Reality

DE GRUYTER

Author
Dr. Hans Peter Good
CH-7320 Sargans

ISBN 978-3-11-061028-4
e-ISBN (PDF) 978-3-11-061238-7
e-ISBN (EPUB) 978-3-11-061048-2

Library of Congress Cataloging-in-Publication Data
Names: Good, Hans Peter, author.
Title: On the origin of natural constants : axiomatic ideas with references to the measurable reality / Hans Peter Good.
Description: Berlin ; Boston : De Gruyter, [2018] | Includes bibliographical references and index.
Identifiers: LCCN 2018032530 (print) | LCCN 2018037671 (ebook) | ISBN 9783110612387 (electronic Portable Document Format (pdf) | ISBN 9783110610284 (print : alk. paper) | ISBN 9783110612387 (e-book pdf) | ISBN 9783110610482 (e-book epub)
Subjects: LCSH: Physical constants. | Physical laws.
Classification: LCC QC39 (ebook) | LCC QC39 .G688 2018 (print) | DDC 530.8/1--dc23
LC record available at https://lccn.loc.gov/2018032530

Bibliographic information published by the Deutsche Nationalbibliothek
The Deutsche Nationalbibliothek lists this publication in the Deutsche Nationalbibliografie; detailed bibliographic data are available on the Internet at http://dnb.dnb.de.

Typesetting: Integra Software Services Pvt. Ltd.
Printing and binding: CPI books GmbH, Leck
Cover image: Science Photo Library/ Kulyk, Mehau

www.degruyter.com

To Graziella

Preface

> Among the greatest discoveries that the human mind has made in recent times belongs the art of judging books without having read them.
> (*Georg Christoph Lichtenberg*)

The natural constants determine the scale of all physical problems and, as quantitative links, form the network of physical theories that do not define the constants themselves. They are determined experimentally as accurately as possible and are the result of a multivariable deviation and equalization calculus of many experiments [V 1]. Since they have been determined from a large number of measurements, they are experimental quantities that have an almost stunning accuracy that far exceeds the ones achieved in a single experiment. Their numerical values and units are not natural laws, but they are based on agreements and are reference points for different physical domains, so that the physical equations are correct. Because of the multivariable equalization calculus, which is monitored by an international committee, the natural constants are very precise, but disadvantageously also model dependent and strongly interlinked. Gravity, for example, calculates with G, relativistic physics assumes a finite speed limit of light in vacuum c, statistical thermodynamics has introduced the Boltzmann constant k_B and quantum physics is determined by the Planck constant h. These consistent building blocks are messages of nature, which according to today's conviction should apply throughout the cosmos. Nevertheless, with a few exceptions, most of their numerical values are useless, because they depend on the system of units chosen and are therefore arbitrary. That is why, in many books, the feeling is often conveyed that natural constants are merely an unavoidable addition of physical laws or a necessary evil. In the best case, they can be found in the form of a table either at the beginning or at the end of a book. In some cases, they are even simply omitted and set equal to one. By this "mathematization", a lot of physics is lost, because among other things dimensional considerations of physical equations are no longer possible.

Is there a connection between the natural constants or are they all just man-made? For what reason do these invariant quantities emerge in all experiments? How many are necessary at all? Is there a mathematical structure behind it? Can a time scale, a length scale as well as a mass scale be chosen arbitrarily? Is the temperature scale completely independent of these three scales? These are basic physical questions that many physicists disagree [V 2]. Such issues are therefore the subject of this book, which attempts to establish a compatible link between different physical scales with an alternative, unified view, and looks beyond the boundaries of a single discipline.

The existence of the natural constants is probably not a coincidence, but the consequence of internal connections. The constant themselves have the greatest potential to tell us something about it. When different natural constants appear in

https://doi.org/10.1515/9783110612387-201

formulas, they inevitably relate separate fields of physics to each other. This structure cannot be inferred or derived mathematically. It has to come from empirical observations by looking for connections between different natural constants. In doing so it should always be kept in mind that the approach is not degenerating into numerical mysticism, that skepticism is appropriate and the statements are compared again and again with experimental data. However, it may well be that there exist some number constants that, detached from the unit system, define some relationships of the different physical scales. Just as the Rydberg or Stefan–Boltzmann constant once lost their irreducible position [V 3], it is conceivable that the Newton constant can also be represented by other constants.

The book begins with the radical interpretation that the mass scale cannot be chosen independent of the time and length scales. Although the unit kilogram is practical for macroscopic phenomena, it describes a derived quantity in the world of units. This perspective arose from the fact that an idea of David Bohm was consistently implemented. He compared the formulas of the electromagnetic energy of a cavity with the energy of a harmonic mass oscillator and found that the formulas for the energy have in both cases an analogous mathematical structure. If it is demanded that corresponding quantities are not only formally but also numerically equal, it follows that the mass unit can be reduced to the dynamic units of length and time. This formal equation, which is elaborated in Chapter 1, is daring, but allows a completely different view of the atomic and subatomic world. It even enables an innovative bridging between gravitation, quantum mechanics and statistical thermodynamics. The axiom derived from the Bohmian analogy is constantly being tested and checked if it matches observations. This postulate is a constant companion throughout the book.

The next three chapters consider universal lengths. The currently available models of physics were first used to search for length scales and to calculate their values from the most fundamental constants of nature. This is based on the idea that fundamental length scales must exist because it is important in physics to know how big things are. By multiplication with the dimensionless Sommerfeld or fine-structure constant (α) as a "geometric factor", the values were then scaled so that they came close to lengths playing a role in the world of physics. The choice of the fine-structure constant may seem arbitrary, but as a scaling factor it is almost mandatory, since it is the only dimensionless quantity given by physics itself. Whyte [V 4] also argued that a length measurement takes a fundamental and unique position in all elementary processes. This view was also shared by W. K. Heisenberg, who pointed out that a mass cannot be calculated from h and c alone, but only along with a fundamental length. Although no lengths occur in Newton's and Maxwell's theories, physics has always been forced to use formulas containing special lengths. The phenomenological description of superconductivity by the London brothers is an example of this and indicates the presence of a constant with the dimension of a length.

Of central importance is the next chapter. It includes the computation of universal quantities using fundamental lengths and the David Bohm analogy. The theory of the metal-insulator transition of N. F. Mott and statistical thermodynamics also play an important role. By simple thermodynamic considerations, a fundamental reference temperature or virtual heat energy can be determined, which is very often referred to in the following chapters. As a consequence, a length comparison offers the possibility of defining the Sommerfeld constant as a number constant. All derived values of the universal quantities are substantiated by experimental examples and measured facts. The idea introduced, which is radically contrary to what physics demands and dictates today, thereby gains in objectivity and, through references to reality, slightly reduces the characteristic of madness. Of course, the view is based on assumptions and is speculative and axiomatic. However, it does refer to a few critical quantities and is an observable-based formalism with no parameters to fit. Without unprovable presuppositions in the form of axioms – with the risk that physical problems may be masked by them – no physical model can be formulated.

From the David Bohm analogy, a fundamental energy density can be derived. This derivation can be found in Chapter 6 and allows that both the mass of the electron m_e as well as the Newton constant G can be defined as quantities, which depend only on the Planck constant h and the speed of light in vacuum c. The latter definition is – similar to the derivation of the fine-structure constant – based on a length comparison. Possibly, this procedure provides an explanation for the fundamental problem, why the electron in the hydrogen atom loses no radiant energy in its ground state despite its motion. By means of the fundamental energy density, it is also possible to define an astronomical length scale, which could be the reason why galaxies are the largest clearly defined building blocks of the universe that form groups. It is not the intention of this book to make predictions. The experimental data should primarily serve to validate assumptions and to "teach an idea running." Nevertheless, the upper limit of the electron neutrino mass is estimated in this chapter, since a corresponding mass of the right order of magnitude can be derived from the fundamental energy density.

Our world seems to be interwoven from the smallest subatomic particle to galaxy clusters. Interestingly, the Bohmian analogy also provides a view pointing in this direction. In Chapter 8, a transformation or duality relation is derived. It converts small lengths into long lengths and vice versa. By means of this transformation, the mean charge radius of the proton is estimated and characteristic energy scales of the nuclear bond are determined. Several thousand nuclear-binding energies can thus be parametrized on average. The MOdified Newtonian Dynamics (MOND) model is explained by this viewpoint as well, and the number of existing fundamental length scales is conclusively increased without new parameters being introduced.

In the next chapter, an additional fundamental length scale is determined by the concept of electrostatic field energy. In addition, stored field energies of homogeneously charged spherical shells with different radii are investigated. If the radius of a spherical shell is replaced by a length developed in previous chapters, it yields to interesting field energies, which can be compared with measured quantities. For example, the calculated relative helium abundance in the Universe matches the measured value very well. It is also surprising that several Raman lines of diamond and graphite can be parameterized with the reference energy and a field energy. Another field energy seems to be the relevant energy scale for the sun's photosphere. Does this energy scale eventually play a role for other stars as well?

The Planck law of heat radiation is undoubtedly one of the most fundamental law of physics. It is reflected in perfect form in the microwave background radiation of our universe. This law is therefore dedicated to a separate chapter. The cosmic microwave temperature T_{CMB} of ≈2.73 K can surprisingly be attributed to a field temperature and by means of a transformation to the reference temperature. The match is near perfect because the calculated value lies in the middle of the error band of T_{CMB}. Today's theories show no relation of T_{CMB} to natural constants. They interpret the microwave temperature as a quantity that can only be determined experimentally. Planck was the first to use the three constants h, c and k_B in the same law. This chapter addresses if the three constants are really independent, since both h and k_B describe the same physical fact of coincidence. Although the background radiation at high energies does not represent Planck's heat radiation, thoughts about radiation at high energies are given in this chapter. It is not surprising that the conversion of nuclear energy into radiant energy that is, the diffuse background radiation at high energies, could be linked to the energy scale of the nuclear binding. It can also be explained with a field energy.

In the following chapter, gravitation, radioactivity and collective magnetism are related. The strengths of these interactions are quantified using dimensionless coupling constants, all of which can be attributed to the fine-structure constant. This quantification succeeds because, due to the David Bohm analogy and the definition of the fine-structure constant as a number constant, both the Newton constant and the electron mass can be represented by the "conversion constants" h and c.[1] These dimensionless coupling constants then allow the calculation of the anomalous magnetic moment of the electron a_e. The anomalous magnetic moment of the electron is the dimensionless physical quantity that has been measured with the greatest accuracy to date. Since all interactions coupled to the electron contribute to this measure, it is the most fundamental reference quantity for physics. It is noteworthy that the parametrization of a_e succeeds with three simple terms, and the value thus calculated within the error limits agrees with the famous precision experiment of the Dyck–Schwinberg–Dehmelt group of the University of Washington. In contrast, according to quantum electrodynamics (QED), the electron is a geometric point without expansion, which interacts with an infinite number of virtual particles. The exchange of the virtual particles is described by means of complex integrals. Is QED really the only possible theory to explain a_e? Definitely, three terms are easier to interpret than

1 The speed of light in vacuum already is a trivial conversion constant in the SI system today.

integrals, which can only be solved with a mathematical trick to avoid infinities. Since the definition of the Sommerfeld constant in Section 5.4 is not bound to a_e, as is the case today, this definition allows an independent verification. Also the description of the anomalous magnetic moment of the muon succeeds with the same number constants. However, the theoretical agreement with the measurement is not given within the error limits.

Despite increasing accuracy, most astronomical measurements cannot be compared to precision measurements in the laboratory. References to such measurement data are therefore subject to greater uncertainties. Nevertheless, in Chapter 12, it is attempted to compute certain astronomical observations, without introducing new parameters, solely with fundamental quantities derived in the previous chapters. This chapter also includes an excursion into the high-energy laboratory of the universe, where millions of times higher energies occur than at the European Organization for Nuclear Research (CERN) with the Large Hadron Collider (LHC). Dimensionless coupling constants could also provide an explanation for the permanent ionizing primary radiation first explored by E. Regener. Similarly, one of the two mysterious kinks in the energy spectrum of cosmic rays can be interpreted.

In Chapter 13, certain observations of the sun and an alternative hypothesis about their mode of operation are discussed by means of fundamental quantities as mentioned in previous chapters.

The next chapter is largely dedicated to experimental particle physics and attempts to explain mass ratios, partial mean lifetimes, magnetic moments of heavy particles and charge radii of the lightest hadrons, without introducing new parameters. Only energy quanta (as detailed in previous chapters), number constants and experimentally confirmed Hall fractions are used. All phenomenological approaches are dimensionless expressions and therefore independent of any measurement system. They also comply with the principle of simplicity and universality. With very few exceptions, all ratios calculated are within the experimental error limits. Overall, approximately 70 data from the Particle Data Group, averaged over several experiments, were analyzed according to the same formalism. The particle masses were based on the reference mass m_e, the charge radii on the Compton length of m_e, the magnetic moments on the Bohr magneton and the average partial lifetimes on the reference time $h/(m_e c^2)$. Independent reference quantities are necessary so that the arbitrariness contained in the definitions of the units has no effect when comparing theory and experiment. For the cataloging of the 70 data, Hall fractions and four dimensionless, physically anchored quantities of previous chapters are sufficient. Remarkably, the same number constants describing particle transformations can be used to calculate the hyperfine splitting of positronium using a simple approach. This chapter is solely an experimental verification of the hypotheses of previous chapters and contains no new findings.

Chapter 15 deals with topics of precision physics. This discipline is the measure of all things to determine the natural constants and the Task Group on Fundamental

Constants of the COmmittee on DATA for Science and Technology (CODATA) extensively references experiments of the simplest atoms that provide the most accurate data by high-resolution spectroscopic methods. Since the publications of the CODATA group contain many formulas and experimental data with references to the original literature in a competent and compact manner, in most cases the CODATA group is directly quoted. When comparing theory with experimental data, only mathematical constants, that is, pure numbers, of previous chapters are used without exception. It is examined whether dimensionless expressions containing only such constants allow a "simple" mathematical description of the data within experimental error limits. The concepts used are neither based on arguments that turn in circles, nor do they involve hidden assumptions. As already practiced in previous chapters, tabulated CODATA values connect the dimensionless expressions to physical reality. This is necessary because concepts without units are mathematically very satisfying, but without standardized "practical units", that is, a classical language, they have no physical meaning.

This book is an explicative attempt to see the construction of matter through a different optic than that of the predominant doctrine. Although every single observation can lead to a discussion, the combined results of many independent tests are not to be dismissed. The conjectures are proved by the fact that they often lead to outstanding agreements with results of experiments that can be verified in the literature. Everyone is free to deduce from the versatile variety of references to reality, whether the conjectures and associations are justifiable or simply wrong. The view cannot be proven and no claim to mathematical rigor is made. It is clear that references to experimental data are not mathematical proofs and that it will be critically questioned in the professional world. The simplicity of the formalism and the evident fact that very often appropriate answers to existing data are obtained from still manageable experiments inspires confidence. However, it does not mean that the point of view is correct. What is certain is that our knowledge is always inadequate and incomplete. Although the assumptions often describe the experimental observations with impressive precision and show phenomenologically in the right direction, skepticism is still appropriate. A questioning is always necessary, whether the right way was taken or not. For it is often possible to find a number of suitable justifications for a result or, as Karl Popper puts it, even if we think that we are coming forward, especially when we think that, we can completely be on the wrong path.[2]

Many physicists are likely to find the whole book suspect and frown, as aspects of recent theoretical models, often justified by their mathematical beauty and elegance, are completely absent. Many things have been worked out intuitively without heavy math by simple merging or by heuristic manipulation with the aid of "non-mathematical" dimensional analysis. Although dimensional analysis is a fundamental principle,

2 Die Quantentheorie und das Schisma der Physik, Mohr Siebeck, page 41.

it probably does not meet the criteria of scientific work in the eyes of some scientists. However, only dimensioned quantities are able to set physical scales and depict reality. The applied principle of orders of magnitude without complicated formalisms lies diametrically in a different direction than today's calculations, which are carried out almost esoterically with ever more powerful computers by taking into account already established principles. The numerical work chosen in this book is simple. It leads to a different physical understanding by means of characteristic length and energy scales than is currently taught in mainstream physics. Unfortunately, today's physics is making things more and more complex by introducing new parameters[3] and using computers. It is indisputable that typical scales must be assigned to specific domains of physics. Only when such scales, where dimensionless factors play a role, are known, do simulations make any sense.

Some physicists certainly agree that current physics is in a dilemma and needs a new conceptual framework because, among other things, mass values of elementary particles cannot be derived from current theories [V 5], [V 6], [V 7], [V 8]. Although the formalism of quantum mechanics is in principle able to explain the structure of atoms with a positive nucleus and negatively charged electrons, it does not provide any indication of the relationship between the absolute values of the speed of light in vacuum (c), the charge (q_e) and the Planck constant (h). Quantum mechanics, like no other theory, drastically reduces the number of free parameters in the physical world. In contrast, the Standard Model of particle physics is overloaded with many parameters in order to explain the observations. The mathematical formalism is enormous and a nonmathematician can hardly understand what is physically going on. To date, there is no method capable of calculating the mass of the proton, the neutron or other particles, although these objects are known as bound states of quarks. Is the physical world so complicated? Or was something overlooked? Hopefully, the considerations at hand provide a small contribution to a more realistic view, as they are based on the integration of natural constants. Physical objects cannot be point particles because of occurring mathematical singularities, but must probably be understood as dynamic geometric structures. The book does not claim to have recognized everything in the description of atomic structures, but only shows that there exist amazingly simple quantitative relationships to which seemingly complicated processes can be attributed. However, many connections are unclear and misunderstood, and a deeper understanding is still needed. The underlying ideas do not emanate from a consistent, mathematically clean theory. Perhaps they are, in particular branches of physics, a hint or a thought-provoking impulse for theoreticians and experts using it for a better formulation or a more mature math. Possibly, it also encourages experimenters in certain branches of physics to make even more precise measurements. It remains

3 The current Standard Model (SM) of particle physics has 27 parameters, all of which must be determined experimentally.

to complete this framework and to deepen the mathematical and the physical understanding as a supplement.

Models and experiments of many scientists made this work possible. The different models are only briefly indicated and not individually described, but only referenced. The details can be found in a wealth of excellent books. All calculations and derivations are carried out only as far as necessary to understand the considerations. There is no claim to mathematical certainty and clarity. Because the math used is very simple and certainly obvious to most, because no group theory, operators and wave functions are used for the description. An attempt is made not to abstract the facts or to mask them with a formal language. What counts is the combination of simple physical models into a consistent whole while preserving geometric principles with length scales that make sense physically. Quite certainly, everything is based on a mathematical structure and symmetry. Probably, number theory and geometry play an important role. If physics is discarded of the units, only arithmetic remains, where the smallest is the natural number one. The postulate of quantum mechanics that energy is quantized requires a discrete math. The deciding factor is, which measure can be attributed to the mathematical symbol of the smallest in the physical reality.

The first part of this book refers solely to the results of published experiments. However, the second part is different where carefully conducted measurements of thin resistive layers are presented that have accumulated over many years in a professional environment. All measurements arose independent of the ideas explained in the first part of the book. The interpretation of these measurements is difficult or impossible with established models. Because in many cases no explanation can be found, such anomalous data usually end up in the bin. Nevertheless, an attempt is made to gain an alternative perspective by means of length and energy scales of the first part. Maybe the measurement results discussed are indeed related to it. Anyhow, it is an exciting experimental journey into a mesoscopic world that can be started independent of the first part of the book. Exciting because these experiments show that in the mesoscopic world the laws of the macrocosm are often no longer valid, because interesting size effects and emergent behavior occur.

References

[V 1] P. J. Mohr. The fundamental constants and theory. Philosophical Transactions of the Royal Society A 363 (2005) 2123–2137

[V 2] M. J. Duff, L. B. Okum and G. Veneziano. Trialogue on the number of fundamental constants. arXiv:physics/0110060

[V 3] E. Hantzsche. Elementary constants of nature. Annalen der Physik 47 (1990) 401–412

[V 4] L. L. Whyte. Über die Eigenschaften einer einheitlichen physikalischen Theorie. Zeitschrift für Physik 56 (1929) 809–817

[V 5] Lee Smolin. The trouble with physics. Mariner Books, September 2007
[V 6] Peter Woit. Not even wrong. Basic Books, New York, September 2006
[V 7] Roger Penrose. The road to reality. Vintage, New York, January 2007
[V 8] Robert B. Laughlin. A Different Universe: Reinventing Physics from the Bottom Down. Basic Books, New York, 2005

Contents

Lists of Figures and Tables

List of Figures

https://doi.org/10.1515/9783110612387-202

List of Tables

Part I

1 The David Bohm analogy: a forgotten idea on the origin of mass

The world is full, the emptiness does not exist.
(*Karl R. Popper, according to the teachings of Parmenides*)[1]

According to classical electrodynamics, an empty cavity filled with electromagnetic radiation has energy and can therefore absorb heat. In thermal equilibrium, the energy density of the electromagnetic fields located in the cavity depends neither on the shape nor on the material of the walls delimiting the cavity. This has been determined experimentally in many cases and is theoretically justified by the fact that the walls only ensure that no energy is exchanged with the environment. It is a completely isolated system with a precise, constant energy that does not change over time. This elementary volume can now be periodically repeated in all three spatial directions. This fills the whole room with elementary cubes of constant energy. With the requirement that the fields within the periodically continued cavities are the same at corresponding points, all conceivable solutions of the Maxwell's equations within the cavity can be represented by a Fourier series. Such a development of the electromagnetic field by a set of discrete Fourier coefficients corresponds figuratively to a decomposition of the field into a set of harmonic oscillators.

It is possible to theoretically express by formulas the electromagnetic energy of an ideal cavity of volume V and the mechanical energy of a harmonic oscillator of mass m. A detailed derivation of these formulas can be found in the book *Quantum Theory* by David Bohm [1.1]. In this book, the author details the transition from classical physics to quantum theory to explain Planck's law of radiation and the Schrödinger equation. According to David Bohm, a comparison of the classical formulas allows a conclusion by analogy of the following form:

Hypothesis 1.1 $m = V / (8\pi c^2)$

The constant c is the speed of light in vacuum, which is currently defined as a fixed quantity. Hypothesis 1.1 results when equations for the electromagnetic energy in a cavity with volume V and the mechanical energy of the harmonic oscillator of mass m have the same mathematical form. In analogy to the relationship mass = energy/c^2, this also allows the view that a mass must correspond to a volume. All systems that oscillate harmonically are quantized, whether they are vibrating matter or electromagnetic waves. If they want to interact with each other, the quantization must be similar for all. To this day, this idea has been confirmed experimentally again and again [1.1].

1 Die Quantentheorie und das Schisma der Physik, Mohr Siebeck, page 191.

https://doi.org/10.1515/9783110612387-001

The derivation of Planck's law of radiation (see also Chapter 10) can be accomplished either from the point of view of the energy spectrum of mechanical oscillators or from the point of view of photons (light quanta) as the smallest unit of radiation energy. The prerequisite is that in both cases the corresponding quantum theoretical generalizations are made. In both cases the wave-particle (matter) duality follows, which plays an important role in quantum theory. The formal similarity between results, obtained from mechanical oscillators that satisfy the Maxwell–Boltzmann relationship, and the phonons of sound[2] that satisfy the Einstein–Bose relationship, may seem random. In any case, both derivations lead to the same result of Planck's radiation law, which excellently describes the experimental data. Probably Hypothesis 1.1 is nothing more than a symmetry requirement of the experimentally observed wave behavior (the behavior of the continuous) and the particle behavior (the behavior of the discrete). In any case, photons and phonons, which are particles of light and sound, respectively are experimentally equivalent and have analogous properties. This has been confirmed thousands of times by large number of experiments. Why do light and sound (matter) follow the same quantum mechanical laws? What is the reason of this symmetry?

David Bohm considered Hypothesis 1.1 to be very important, which is derived by comparing the electromagnetic energy in a cavity with the energy of a mechanical oscillator. But he elaborated the explicitly written analogy (Hypothesis 1.1) for the mass no further. However, this is exactly what will be developed in the following chapters. Hypothesis 1.1 offers the possibility of integrating a fundamental mass scale into quantum theory and, though daring and very speculative, approximating gravitation and quantum mechanics on a trial basis. If the formal analogy with the harmonic matter oscillator established by David Bohm is thought through to the end, and the mass is understood as a "quantum mechanical quantity" that arises through motion, then for the units it follows:

Relation 1.1 (mass) = (length) $(\text{time})^2$; in SI units:kg = m s^2

By Relation 1.1, the unit of mass is connected by a simple linear relationship with the unit of length and the square of the unit of time as a proportionality factor. This reduces the dimensional space by 1. The fact that kg and m s^2 are dimensionally interconvertible eliminates the unit of the kilogram when the units m and s are given a fundamental status. As coulomb is a product of the ampere and the second, kilogram is a product of the meter and the square of the second.

2 The existence of these particles is coupled to macroscopic matter. When the crystalline solid is decomposed into atoms, the particles also disappear.

From Relation 1.1, Relation 1.2 between SI units is derived as follows:

Relation 1.2 $\quad \text{kg m}^{-1}\text{ s}^{-2} = \text{Pa} = \text{J m}^{-3} = 1$

The replacement of the unit kg by the composite unit m s^2 obviously makes the unit of energy density and pressure dimensionless. This means that energy density and pressure must have a fundamental physical meaning, since they become independent of any man-made units because of Relation 1.2.

In the current SI system, of the three units, kg, m and s, only units kg and s are independent of each other, because meter is fixed by second and speed of light in vacuum. If time is converted to length with the help of the speed of light in vacuum, mass has the unit m^3/c^2, that is, volume per c^2, as it must be according to Hypothesis 1.1.

The utilization of Einstein's statement that energy and relativistic mass are linked together by the square of the speed of light, that is, energy can become matter or matter can become energy, combines together with Hypothesis 1.1 to a finite relativistic energy density of a ground substance of space of $(mc^2)/V = (8\pi)^{-1}$ J m^{-3}. This basic substance is convertible and can appear as a stable or a transient particle, for example, as an electron or a pion.

With Relation 1.1 it is possible to reduce the unit of mass to other constants of nature, and to replace the primary kilogram as a reference object. While this does not satisfy metrologists' requirement that a redefinition must be based on commonly accepted physical laws, the number of fundamental constants is reduced. Such a procedure offers enormous advantages, since the kilogram is the only SI base unit that is still realized by a prototype body and is therefore neither independently defined by location nor time . Although there are several prototype bodies in different countries, they are steadily losing mass, creating differences that are difficult to control.

Each unit of time can be converted to a length or vice versa. Similarly, each length is equivalent to an inverse mass. This is due to the fact that time can be rotated in space in a mathematically well defined manner. However, it is difficult to see why Relation 1.1 holds, and the unit kg and the unit m s^2 are manifestations of the same physical reality. Mass cannot be destroyed, it is only converted – in analogy to Einstein's equivalence formula – into something else with another derived unit. As pressure is created by a collective motion of many particles, the mass is an emergent consequence of small-scale dynamics. The relativistic mass is dynamically generated at the level of quanta, and can be traced back to a dynamic relationship between space and time, is decidedly a great assertion whose correctness can be called into question. It is understandable that it cannot be readily accepted that the unit kg and the unit m s^2 are interchangeable terms with a proportionality factor of 1, and the physical doctrine must be omitted. Since every mathematical deduction of the term mass on a microscopic level has failed to date, and it is still unclear how mass could be reduced to natural constants, this axiomatic approach will be used to further analyze its physical consequences.

The fact that the vital unit kg is only a "convenient" unit that can be replaced by m s^2 is probably irritating and shocking for most physicists, and some therefore stop reading. There is no need indeed to believe this, because this principle, which must be absolute, is not readily observable. Certainly, many questions are raised when the daring Relation 1.1 serves as a speculative idea and mathematical tool[3] for the following chapters, and it is claimed that physics actually does not need the kilogram. Only references to experiments can convincingly clarify whether such a perspective is useful for the understanding of physics or whether the hypothesis is an idle aberration. If this "heuristic concept" has a justification for the description of the physical reality, it should be observable everywhere, and various phenomena in different parts of physics should be explained as a consequence of Relation 1.1. In any case, from a physical point of view, the rationale and working hypothesis, formulated by means of Relation 1.1, is the guiding principle and the Achilles heel of all the chapters of Part I.

Reference

[1.1] David Bohm. Quantum Theory. Prentice-Hall, Inc., Englewood Cliffs, N. J., 1951, p. 14

3 In quantum mechanics, it is still unclear why measurement results can theoretically be described if the variables of the classical Hamilton function are transformed into operators by means of rules.

2 Universal lengths of atomic physics and statistical physics

Si toute particule qui nous apparaît comme isolée à l'échelle microscopique peut constamment échanger de l'énergie et de la quantité de mouvement avec le milieu subquantique, celui-ci joue le rôle d'un « thermostat caché » avec lequel la particule est constamment en contact énergétique. (*Louis De Broglie*)[1]

Negatively charged electrons can be removed from bonds by light or by applying a voltage, and then these electrons can move freely. The resulting positively charged defects, called holes, are also free to move and behave in a similar way as free particles, which are not actually particles, but behave as if they are. The properties of holes (quasi-particles) are analogous to the properties of electrons added to the material, but with a positive charge. These vacancies or states of negative energy are unoccupied states or occupational possibilities, which are not occupied by the electrons. Holes can form electron-bound electron-hole pairs, which carry energy but no charge because they are electrically neutral. Louis De Broglie suggested that such holes in a sea of particles (Dirac sea) could be called anti-particles. Presently, it is believed that the analogy between anti-particles, voids or holes in materials is universal.

For bound particles, the average Coulomb energy in vacuum (E_B) between a hole (positron) and an electron is given by the following formula:

Formula 2.1 $E_B = q_e^2\,(4\pi\varepsilon_0)^{-1}\,(2a_B)^{-1}$

where a_B denotes the relevant Bohr radius, which describes the quantum mechanical expansion of the wave function in the ground state, that is, the size of the electron-hole pair. The quantity $2a_B$ can also be understood as a "bond length", which determines the concentration of particles in the sea of particles. In order to release a hole or an electron , the energy amount E_B must be applied as ionization energy.

The abovementioned model stems from atomic physics and can be derived using the nonrelativistic Bohr model. If, for example, the Bohr radius of the hydrogen atom $a_H \approx 0.529$ Å is used for the relevant Bohr radius a_B, the ionization energy of hydrogen of ≈ 13.6 eV results. The observed ionization energy of helium, which has the highest ionization energy of all elements, is ≈ 24.6 eV. From this, a relevant Bohr radius of helium of ≈ 0.293 Å can be calculated, which is interestingly about 5/9 of the radius of the hydrogen atom a_H. The number 9 in simple fractions will play an important role in the following chapters. It may therefore not be a coincidence that the number 5/9 appears in this context. Helium is the second most abundant element of the observable

1 La thermodynamique « cachée » des particules. Annales de l'Institut Henri Poincaré A1 (1964) 1–19

https://doi.org/10.1515/9783110612387-002

universe besides hydrogen. Formula 2.1 involves the idea that physically comparable systems behave similarly and energy is heuristically always equivalent to length.

If it is assumed that a particle with mass m_0 is responsible for binding (interaction), that is, the mass of a free particle is transformed into energy, which is available as binding energy, then

Hypothesis 2.1 $m_0c^2 = E_B$

must apply because of the equivalence of mass and energy according to the special theory[2] of Albert Einstein. Assuming Hypothesis 2.1, the binding energy E_B is equated to the virtual mass m_0. If positive and negative electrons already exist, it may not be so absurd to postulate a third neutral particle m_0 associated with the positron and the electron. Basically, this means that from an electrically neutral electron-hole pair, a neutral particle with a rest energy can arise, which corresponds to the ionization energy of the electron-hole pair. To describe the extent of the virtual particle of mass m_0, its Compton length Λ can be used. Thus, the particle of mass m_0 remains localized to a region of space required for the binding. According to Louis De Broglie

Definition 2.1 $\Lambda = h\, c^{-1}\, m_0^{-1}$

must apply. If E_B from Formula 2.1 and m_0 from Definition 2.1 are combined in Hypothesis 2.1, then the following definition can be derived:

Definition 2.2 $a_B = (4\pi)^{-1}\, \alpha\, \Lambda$

where α is the famous Sommerfeld constant $\alpha \equiv q_e^2/(4\pi\varepsilon_0)/(h_{bar}c)$ originally introduced by Arnold Sommerfeld. He introduced this dimensionless constant almost a century ago to explain the fine structure of spectral lines of hydrogen. For this reason, this quantity is often called fine-structure constant. In recent physics, this constant determines the strength of the force between electrically charged particles. Because of natural constants occurring in the expression, it is a synthesis of quantum theory (h), electrodynamics (q_e, ε_0) and relativity (c). Sommerfeld's constant is the same in every system of units and it is determined as a scaling factor in energy scales of atomic physics in the unit m_ec^2. Due to its fundamental properties, its theoretical explanation is the central problem of natural philosophy.

In statistical physics, it is believed that a system of particles formed by microstates can be described by a few macroscopic parameters. In this theory, the concept of the thermal De Broglie length λ_0 proves to be very useful. For a noninteracting,

2 From a physical point of view, the special theory of Albert Einstein is unsatisfactory because the uniform motion is preferred in the formulation of the theory.

ultra-relativistic particle in equilibrium with a heat bath of temperature T, the thermal De Broglie length λ_0 is given [2.1] by

Formula 2.2 $\lambda_0 = h\, c\, 2^{-1}\, \pi^{-1/3}\, (k_B T)^{-1}$

This applies on the premise that the energy (ε) and the momentum (p) of a massless particle are linked by the relationship $\varepsilon = cp$. If the temperature T is replaced by a "reference temperature" τ, and it is axiomatically demanded that the thermal De Broglie length λ_0 corresponds to the Compton length Λ of the virtual mass m_0, then

Hypothesis 2.2 $\Lambda = \lambda_0 = h\, c\, 2^{-1}\, \pi^{-1/3}\, (k_B \tau)^{-1}$

results. The interaction between a hole, an electron and an unbound particle with mass m_0 is thus understood as a relativistic, thermodynamic system with the length scale Λ and the statistically defined reference temperature τ. If $E_\tau = k_B \tau$ is introduced as a "relativistic" reference energy, it is possible to write the following for the length Λ, a_B and the mass m_0:

Relation 2.1 $\Lambda(E_\tau) = h\, c\, 2^{-1}\, \pi^{-1/3}\, E_\tau^{-1}$

Relation 2.2 $a_B(E_\tau) = \alpha\, (4\pi)^{-1}\, \Lambda(E_\tau)$

Relation 2.3 $m_0(E_\tau) = h\, c^{-1}\, \Lambda(E_\tau)^{-1}$

Thus, the quantities Λ, a_B and m_0 are functions of the as yet unknown "relativistic" reference energy E_τ or what is equivalent, the reference temperature τ, when the two quantities E_τ and τ are formally linked by the Boltzmann constant k_B. With Relation 2.1 to 2.3, a simple formulation with few parameters for the relativistic interaction of three particles is achieved by principles of statistical mechanics. The introduction of the empirical quantity τ eliminates the need to reduce the interaction solely to individual particles, and the tracking and characterization of the particles become obsolete. It achieves a dramatic simplification of the microscopic random dynamics (chaos) of the many-body system by means of thermodynamic movements. Since in addition to the reference temperature τ, only natural constants occur in the expressions, and it is easily possible to calculate the quantities Λ, a_B and m_0 when the reference temperature τ is known. But how can the reference energy E_τ be calculated? This will be the central theme in Chapter 5.

Reference

[2.1] Zijun Yan. General thermal wavelength and its applications. European Journal of Physics 21 (2000) 625–631

3 Charge localization and delocalization

The researcher has to listen to nature's general principles by looking at certain general features in larger complexes of empirical facts.
(*Albert Einstein*)[1]

In 1927, Karl F. Herzfeld [3.1] was the first to quantify what makes a material a metal or an insulator without the use of quantum mechanics. Both the term metal and the term insulator can only be defined macroscopically, because in the microcosm these concepts do not make any sense. Karl F. Herzfeld used the Clausius–Mossotti relationship, which correlates the macroscopic permittivity (ε_r) and the macroscopic electrical susceptibility (χ_e) with the microscopic electrical polarizability (α_{pol}) of a nonpolar free particle of number density (n). The answer is as follows:

Formula 3.1 $\varepsilon_r = 1+\chi_e = 1+3\,n\,\alpha_{pol}\,(3\,\varepsilon_0 - n\,\alpha_{pol})^{-1}$

For the critical condition (Condition 3.1),

Condition 3.1 $n\,\alpha_{pol} = 3\,\varepsilon_0$

the static permittivity ε_r of Formula 3.1 diverges. The instability causes the bound valence electrons to be spontaneously released, and to form a metallic conductor. In other words, it destroys the localization of electrons and allows the expansion of electrons over longer ranges. This singular point or unstable state in which metallic behavior sets in is also called dielectric catastrophe or Goldhammer–Herzfeld instability. At this point, with sufficiently strong interaction, the local dipoles spontaneously polarize and form phases of particular orientation.

When a hydrogen atom is brought into an external electric field, the proton and the electron shift relative to each other and a dipole moment proportional to the electric field are induced. By applying perturbation theory to the Schrödinger equation, which is one of the most fundamental microscopic equations of physics,

Formula 3.2 $\alpha_{pol} = (9/2)\,(4\pi\varepsilon_0)\,a_H^3$

results for the microscopic polarizability of the hydrogen atom [3.2]. The quantity $\alpha_{pol}/(4\pi\varepsilon_0)$ is called the polarizability volume (unit m^3). Assuming that this calculation is valid for all hydrogen-like systems, the relevant Bohr radius a_H^* must be used for a_H. Combining Formula 3.2 and the critical Condition 3.1 gives the following relation:

1 Sitzungsberichte der Preussischen Akademie, Antrittsrede des Hrn.Einstein, 1914, Seite 740.

https://doi.org/10.1515/9783110612387-003

Relation 3.1 $n^{1/3} a_H^* = (6\pi)^{-1/3} \approx 0.3757$

Since $n^{-1/3}$ corresponds to the mean distance of the particles, $n^{1/3} a_H^*$ can be regarded as the ratio of the size of a bound (localized) particle to the mean distance between free particles. This means that in a space element of size a_H^{*3} there must be about 0.3757^3 (≈ 0.053) particles.

3.1 Metallic liquids

The metal-insulator transition both heuristically and theoretically has been studied since a very long time. Nevertheless, it still causes lively discussions. In reference [3.3] the different properties of physics of this phenomenon are compiled. It was postulated very early that all matter, if strongly compressed, should have metallic properties. At high pressure, the atoms should come so close that the outermost electrons can move freely between the atomic cores, and in extreme cases, matter forms plasma that no longer has atoms but only freely movable charge carriers. It is believed that the interior of the gas planets Jupiter and Saturn fulfils this condition, and the objects therefore consist of liquid metallic hydrogen because of the pressure prevailing there.

It has been found experimentally [3.4], [3.5], [3.6] that the elements hydrogen (H), nitrogen (N), oxygen (O), rubidium (Rb) and cesium (Cs) behave comparably. These fluids even show that density and temperature are even more important than the pressure itself at the critical transition. The temperature required for transition is in all cases ≈ 2,000 K. At this transition, that is, when the ratio of the size of an isolated atom (a_H^*) to the mean distance between the atoms ($n^{-1/3}$) reaches a critical value, all five elements have a limiting metallic conductivity of ≈ 2,000 Ω^{-1} cm^{-1}. For H, Cs and Rb, the experimental value of the scaling parameter $n^{1/3} a_H^*$ is ≈ 0.38; for nitrogen and oxygen it is ≈ 0.34. These values are quite close to the value of Relation 3.1 and allow us to conclude that Relation 3.1 must be fundamentally correct [3.6], because experimental results can thus adequately be described. Metallic liquids show that it is not possible to determine whether a metal or a nonmetal is present if direct-current conductivity is measured or not measured at very low temperatures.

3.2 The universal screening length and the Mott metal-insulator transition

A positively charged impurity in a conductor causes a disturbance in the electron concentration in its vicinity, which shields the electric field of the impurity. The positive charge of the impurity will attract electrons and repel positive charges. Thus a negatively charged cloud is formed around the impurity within a spherical shell, which reduces the electric field caused by the impurity. However, full neutralization of the impurity cannot occur without energy input and the material remains polarized. The relevant physical quantity for the shielding described above is the Thomas–Fermi shielding length λ_{TF} given by the following formula:

Formula 3.3 $\lambda_{TF}^2 = (\varepsilon_0/q_e^2)\, D^{-1}$

D denotes a double density of states with the unit $J^{-1}\, m^{-3}$ and specifies the number of quantum states per unit energy and per unit volume. The Thomas–Fermi shielding length is the distance over which the ordinary Coulomb field is exponentially attenuated by polarization of the material, thereby confining the Coulomb field to a narrow space. For a spherical symmetry, the Thomas–Fermi potential is given by the following formula:

Formula 3.4 $\Phi(r) = q_e\, (4\pi\varepsilon_0\, r)^{-1} \exp\{-r/\lambda_{TF}\}$

The Thomas–Fermi potential is obtained by solving the Poisson equation. The Thomas–Fermi shielding is a collective effect of charge carriers and is caused by long-range electrostatic fields. Above a critical electron concentration, the shielding length becomes so small that the valence electrons can no longer be bound and metallic behavior results. Below this critical concentration, that is, at low electron concentrations, the shielding length is large, the valence electrons are bound and the material becomes an insulator. The critical point, that is, the boundary condition for binding or no binding to the impurity, was estimated by N. F. Mott [3.7] by using the condition $\lambda_{TF} = a_H^*$. He employed for the density of states D in Formula 3.3 the density of states of the Fermi energy D_F of the free electron gas, which is given by the following formula:

Formula 3.5 $D_F = (3/2)\, n\, (k_B T_F)^{-1}$

Combining Formula 3.5 with the definition of the Fermi temperature of the free electron gas T_F yields the following relation:

Relation 3.2 $n^{1/3}\, a_H^* \approx 0.25$

Relation 3.2 is also called the Mott criterion. N. F. Mott was the first to derive the criterion (Relation 3.2) based on a thermodynamic principle, and to correlate the relevant Bohr radius a_H^* with the critical number density n of the electron-hole pairs at the transition. There is a wealth of literature on both the experimental determination of the particle number density n and the theoretical calculation of the right-hand constant of Relation 3.2.

The model of a charged impurity in a conductor can be analogously transferred to virtual photons in vacuum. A virtual photon can split into a virtual electron-positron pair for a very short time and thereby "polarize" the vacuum with different charges (vacuum polarization). Each charge is surrounded by a shell of electron-positron pairs, which shield the charge to greater distances. The effect occurs because virtual charges in the vacuum arise and disappear dynamically.

As a working hypothesis, the constant of the right side of Relation 3.2 shall now be calculated in a different way. If the Fermi temperature T_F is replaced by the reference temperature τ defined by Hypothesis 2.2, and $2^{1/2}$ instead of 3/2 is chosen as the proportionality factor, then Formula 3.5 can be written as the following:

Hypothesis 3.1 $D = 2^{1/2}\, n\, (k_B \tau)^{-1}$

Using Formula 3.3 together with Hypothesis 3.1,

Relation 3.3 $n^{1/3}\, a_B = (4\, \pi^{4/3})^{-1/3}$

remains if $\lambda_{TF} = a_B$ is additionally postulated. For simplicity, the constant c_1 is defined as follows:

Definition 3.1 $c_1 \equiv (4\, \pi^{4/3})^{-1/3} \approx 0.3787$

It is astonishing how well the constant c_1, calculated by means of the quantum state density of Hypothesis 3.1, agrees with the experiment [3.4] and the value calculated with the Schrödinger equation (≈ 0.38). In particular, the quantitative agreement with the value obtained by perturbation theory from the linear Schrödinger equation (Relation 3.1) is remarkable. Is this an indication of the correctness of the analogy conclusion and of the proper selection of the arbitrarily chosen proportionality factor $2^{1/2}$ in Hypothesis 3.1? In any case, even if a slightly modified assumption has been made for the density of states by Hypothesis 3.1, the computation of the constant c_1 is based on similar physical arguments that N. F. Mott used to derive his value.

In plasma physics, the dimensionless Debye number N_{Debye}, that is, the mean number of charge carriers within a sphere of radius of the screening length (Debye sphere) plays a central role. Setting $\lambda_{TF} = a_B$, this combined together with Relation 3.3 gives the following relation:

Relation 3.4 $N_{Debye} = (4\pi/3)\, n\, a_B{}^3 = 3^{-1}\, \pi^{-1/3} \approx 0.228$

3.3 The nine-dimensional hypersphere as a possible ordering principle and the meaning of π

For the description of the Mott metal-insulator transition, the constant c_1 in equation $n^{1/3}\, a_H{}^* = c_1$ is empirically calculated in this chapter as an alternative to existing models. Interestingly, despite some empirical evidence, the constant c_1 has an additional geometrical property that is not obvious at first glance.[2] This property appears to be reflected in the periodic table and the most intense spectral lines of the hydrogen atom as well.

Let V_m be the volume of an m-dimensional hypersphere with radius r. Then the reciprocal hypervolume V_m^{-1} is given by [3.8]

Formula 3.6 $V_m^{-1}(r) = \Gamma(1+m/2)\,\pi^{-m/2}\,r^{-m}$

For r = 1, volumes of m-dimensional unit spheres result. For example, $V_1(1) = 2$ and $V_2(1) = \pi$. Does the number 2, similar to π, have its origin in geometry?

Let the radius of the hypersphere be determined by the constant c_1 according to

Condition 3.2 $r = c_1 = (4\,\pi^{4/3})^{-1/3}$

It can be shown analytically that only for the odd dimension m = 9, the number π for V_m^{-1} vanishes if the constant c_1 is chosen for the radius. The reciprocal hyper volume $V_9^{-1}(c_1)$ is then given by the following Relation 3.5:

Relation 3.5 $V_9^{-1}(c_1) = 1890 = 2\cdot3^3\cdot5\cdot7$

In Relation 3.5 the prime number decomposition of the whole number 1890 is also given. Because of the fundamental theorem of arithmetic, any natural number can be decomposed into prime factors. It is therefore not unusual that the number 1890 can be prime factorized as well. It is interesting, however, that only the four smallest primes 2, 3, 5 and 7 arise in the decomposition. It seems that the constant c_1 particularly distinguishes the volume of the nine-dimensional hypersphere. What role does this volume play in the mathematical description of nature? What is the significance of hyper space geometry in physics in general? In mathematics, the number 9 is not an unknown: In 9 dimensions, the central hypersphere touches all 18 confining side surfaces of the outer hyper cube. The number 9, according to the assumption made by Edward Waring, is the minimum number that is necessary for any natural number to be represented either as a cubic number or as a sum of 2,3,4,5,6,7,8 or 9 cubic numbers.

The Debye number N_{Debye} computed with Relation 3.4 can be attributed to the volume of a hypersphere. It applies the following relation:

Relation 3.6 $N_{Debye} = 3^{-1}\,\pi^{-1/3} = V_3(c_1)$

2 After an idea of David Kummer.

3.3.1 The mass ratio m_p/m_e

The famous dimensionless constant, namely the mass ratio m_p/m_e of proton and electron, has often stimulated speculations and gave rise to ingenious hypotheses. This mass ratio, like the Sommerfeld constant, is one of the most important dimensionless constants of physics, and nevertheless it is one of the greatest puzzles of physics. Although it is known from experiments how large the mass ratio is, there is still no physical theory to calculate its value. Why is the mass ratio m_p/m_e of proton and electron, which is about 1836 and has so far escaped all understanding, on the order of 1890? Both fine-structure constant and m_p/m_e are independent of any system of units and imply that there must be a connection to a mathematical scheme. The question is which. In any case, a property of the mathematical structure of $V_9^{-1}(c_1)$ seems to be roughly reproduced in the mass ratio m_p/m_e. A hint worth thinking about more seriously and looking at in more detail?

Table 3.1: Classification of the elements.

Period (shell)	Configuration in today's spelling	Number of elements or number of electrons in a period		
1	1s	$2\cdot(1)$	$=2\cdot1^2$	=2
2	[He] 2s 2p	$2\cdot(1+3)$	$=2\cdot2^2$	=8
3	[Ne] 3s 3p	$2\cdot(1+3)$		=8
4	[Ar] 3d 4s 4p	$2\cdot(1+3+5)$	$=2\cdot3^2$	=18
5	[Kr] 4d 5s 5p	$2\cdot(1+3+5)$		=18
6	[Xe] 4f 5d 6s 6p	$2\cdot(1+3+5+7)$	$=2\cdot4^2$	=32
7	[Xe] 4f 5d 6s	$2\cdot(1+3+5+7)$		=32
8	[Rn] 5f 6d 7s	$2\cdot(1+3+5+7+9)$	$=2\cdot5^2$	=50

Note: A horizontal row of the periodic table is called period, which has a very specific number of elements. The chemically inert noble gases are at the end of each period.

3.3.2 The periodic table of elements

Matter is made up of a limited number of simple substances called elements. These substances cannot chemically be decomposed into other substances. Through experiments, it has been found that the chemical behavior of the elements in the electrically neutral state changes periodically, that is, it repeats after 2, 8, 18 and 32 elements. The place in the periodic table is determined by the number of protons in the nucleus.

The periodic table has 118 (=2 + 8 + 8 + 18 + 18 + 32 + 32) elements. Table 3.1 summarizes the atomic systematics in a clear and compact way using the prime numbers 2, 3, 5 and 7. This table also lists the atomic configuration that results from Bohr's model. Is it not surprising that the periodic table can be interpreted with the prime factors of 1890? Why do these numbers play a role in the systematics of the periodic table? What is the principle behind it?

3.3.3 Stability of atomic nuclei

Nuclei consist of neutrons and protons, which are held together by the strong interaction. If the separation energy is considered, that is, the energy to remove a proton or neutron from the nucleus, it is empirically found that extremely high values exist when the number of protons or neutrons is 2(He), 8(O), 14(Si), 20(Ca), 28(Ni), 50(Sn) or 82(Pb). These isotopes are much more stable than those that have more or less nucleons, and they are particularly abundant in the cosmos. “Double magic” nuclei, that is, nuclei with magic numbers of protons and neutrons are particularly stable. As shown in Table 3.2, the number of protons or neutrons of these “magic nuclei” can all be represented with one exception by products of prime factors of 1890.

Table 3.2: Magic nuclei.

Number of protons or neutrons	Element	Product representation
2	He	2^1
8	O	2^3
(14)	Si	$2^1 \cdot 7$
20	Ca	$2^2 \cdot 5$
28	Ni	$2^2 \cdot 7$
50	Sn	$5^2 \cdot 2$
82	Pb	no explanation
126 (only for neutrons)		$2 \cdot 3^3 \cdot 7$

Note: The so-called magic nuclei, which are more stable than those that have more or less nucleons.

The element lead with atomic number 82 is the last element that is not radioactive. The next element bismuth with atomic number 83 has an extremely long radioactive half-life. All subsequent elements are unstable and decay without exception. Within the 83 elements there is technetium with the atomic number 43 (a prime number) and promethium with the atomic number 61 (a prime number), which do not occur in nature. So there are totally 81 ($=9^2$) stable elements when the heaviest element bismuth is also counted. Molten bismuth, like water, expands on solidification (density anomaly), and is the strongest diamagnetic element among all elements. Interestingly, the difference in the atomic numbers of promethium and technetium is a multiple of 9. The number of unstable elements occurring in nature is also 9. These are the elements with atomic numbers $84 \leq Z \leq 92$. Are these observations all coincidences?

Table 3.3: Emission lines of atomic hydrogen.

$\Lambda_{observed}$/Å	Relative intensity		Name	Prime number product	Decimal value	$\lambda_{obs} / \lambda_{H\gamma}$
972.517	83000	$5^2 \cdot 3^{-3}$	Ly-γ	$2^5 \cdot 7$	224	0.223 991
1215.6701	840000	$2^2 \cdot 7 \cdot 3^{-1}$	Ly-α	$2^3 \cdot 5 \cdot 7$	280	0.279 994
3889.064	70000	$7 \cdot 3^{-2}$	Hζ	$2^7 \cdot 7$	896	0.896 000 6
4101.734	70000	$7 \cdot 3^{-2}$	Hδ	$3^3 \cdot 5 \cdot 7$	945	0.944 997
4340.472	90000	1	Hγ	$2^3 \cdot 5^3$	1000	1
4861.35	180000	2	Hβ	$2^5 \cdot 5 \cdot 7$	1120	0.1120 005
6562.79	500000	$2 \cdot 5^2 \cdot 3^{-2}$	Hα	$2^3 \cdot 3^3 \cdot 7$	1512	0.1511 999

Experimental data: Yu. Ralchenko et al. [3.9]
Prime number products: http://www.mathpages.com [3.10]

Note: Prime number products, consisting of powers of prime numbers 2, 3, 5 and 7, can be assigned to the seven most intense emission lines λ_{obs} of atomic hydrogen. Normalization of λ_{obs} with $\lambda_{H\gamma}$ was carried out to get simple decimals for $\lambda_{obs} / \lambda_{H\gamma}$, and to make the deviations better visible. Since the Lyman lines refer to vacuum and the Balmer lines to air, the ratios $\lambda_{obs} / \lambda_{H\gamma}$ for the Lyman lines were divided by 1.0003. It is surprising also for relative intensities, about which the Bohr model makes no statements, can be represented by prime number products.

3.3.4 The strongest emission lines of atomic hydrogen

To explain the optical emission lines of atomic hydrogen, the semiclassical Bohr model is usually used. The agreement of the experimental data with the very simple Bohr model is undoubtedly excellent and, in principle, invalidates another interpretation. Based on the smallest primes, Table 3.3 shows an additional interpretation that is not accessible by the Bohr model. Interestingly, all strong emission lines of atomic hydrogen can also be represented as products of the primes 2, 3, 5 and 7. Most of the lower intensity emission lines are not attainable in this way, though. To what extent do the four smallest primes 2, 3, 5 and 7 control the behavior of our reality? Why is the whole periodic table so interlocked with $V_9^{-1}(c_1)$?

References

[3.1] K. F. Herzfeld. On atomic properties which make an element a metal. Physical Review 29 (1927) 701–705

[3.2] L. I. Schiff. Quantum mechanics. MacGraw-Hill Inc. (1968), section 33

[3.3] P. P. Edwards, R. L. Johnston, C. N. R. Rao, D. P. Tunstall and F. Hensel. The metal-insulator transition: a perspective. Philosophical Transactions of the Royal Society A 356 (1998) 5–22

[3.4] W. J. Nellis. Universal behaviour of nonmetal-metal Mott transitions in fluid H, N, O, Rb and Cs. Journal of Physics: Condensed Matter 16 (2004) 923–928

[3.5] F. Hensel and P. P. Edwards. Hydrogen, the first alkali metal. Chemistry – A European Journal 2 (1996) 1201–1203

[3.6] W. J. Nellis, R. Chau, P. P. Edwards and R. Winter. The transition to the metallic state in alkali and low-z fluids. Zeitschrift für physikalische Chemie 217 (2003) 795–802

[3.7] N. F. Mott. The transition to the metallic state. Philosophical Magazine 6 (1961) 287–307

[3.8] M. G. Kendall, Sc.D. A course in geometry of n dimensions. Charles Griffin & Company Limited London

[3.9] Ralchenko, Yu., Kramida, A.E., Reader, J., and NIST ASD Team (2011). NIST Atomic Spectra Database (ver. 4.1.0), [2012, April 19], National Institute of Standards and Technology. http://physics.nist.gov/asd

[3.10] http://www.mathpages.com, queried on October 19, 2013

4 The universal length for short-range order

The similarities between the vacuum of space and low-temperature phases of matter are legendary in physics.
(*Robert B. Laughlin*)[1]

The Mott metal-insulator transition of metallic fluids has been instrumental in experimentally establishing the constant c_1. Since phase transitions based on quantum properties of matter are of great importance, this chapter will explain further properties and experimental findings of such transitions. The viewpoint is based on the idea that in many cases fundamental phase transformations can be described solely by natural constants.

The charge transport in metals can classically be described by the Drude model. In quantum mechanics, the charge transport is modeled by a free electron gas in three dimensions. By combining these two models (quasi-classical theory), the specific electrical resistance ρ_{el} can be written as follows [4.1]:

Formula 4.1 $\rho_{el}^{-1} = (2/3)\,\pi^{-1}\,q_e^2\,h^{-1}\,(k_F\,l_e)\,k_F$

Electrons can only be localized (isolated) if the electron waves with wavelength $\lambda_F = 2\pi/k_F$ are scattered within their coherence length l_e. That is, in a strongly scattering medium, the wave function is limited to a volume l_e^3 determined by the mean free path, and an electron cannot perform a single oscillation before it is scattered again. At the critical point, a pronounced loss of coherence occurs and

Condition 4.1 $k_F\,l_e = 1$

according to Ioffe and Regel applies [4.2]. This limiting condition for the localization gives the following relation:

Relation 4.1 $\rho_{el} = (3/4)\,(h/q_e^2)\,\lambda_F \approx 194\,(\lambda_F/\text{Å})\;\mu\Omega\,\text{cm}$

This relation designates the resistivity at the critical point, that is, at the metal-insulator (MI) transition. All metallic fluids described in Chapter 3 show saturation of the electrical resistivity at $\approx 500\;\mu\Omega$ cm. With this value, a wavelength $\lambda_F \approx 2.6$ Å results from Relation 4.1. Although all metallic liquids are chemically different, a similar value is obtained for λ_F.[2] This raises the question whether the wavelength λ_F

1 A Different Universe, Basic Books (2005), page 105.

2 Covalent bond distances of most metals, that is, metal-metal bond lengths of homonuclear single bonds, are between 2 and 3 Å.

https://doi.org/10.1515/9783110612387-004

is a universal quantity or not. If the length λ_F is universal, it must be representable by natural constants. Let the universal length λ_{MI} be fixed by the following definition:

Definition 4.1 $\lambda_{MI} \equiv \varepsilon_0^{-1}\, c^{-2}\, q_e^2\, m_e^{-1}\, \alpha^{-2} = 2\, h\, c^{-1}\, \alpha^{-1}\, m_e^{-1} = 2\, \alpha^{-1}\, \lambda_e \approx 6.65\ \text{Å}$

If an attempt is made to set λ_F equal to λ_{MI} in Relation 4.1, this gives a specific resistance at the metal-insulator transition, which is too large. Something does not seem to fit.

With the phenomenological Ansatz formula

Ansatz 4.1 $\rho_{el} = (h/q_e^2)\, \lambda_F\, (k_F\, l_e)^{-1} = R_K\, \lambda_F\, (k_F\, l_e)^{-1} \approx 258\, (\lambda_F/\text{Å})\, (k_F\, l_e)^{-1}\ \mu\Omega\ \text{cm}$

a different interpretation of the resistivity is obtained. The quantity R_K is the von Klitzing constant measured in the unit Ohm. It is the quantum of the quantum Hall effect and is named after its discoverer Klaus von Klitzing. Its value is ≈ 25 812.807 Ω. R_K is related to Planck's constant h and the electron charge q_e through the relationship $R_K = h/q_e^2$ and is now used as the unit of electrical resistance. Setting $(k_F\, l_e)_{MI} = 7/4$ and $\lambda_F = \lambda_{MI} \approx 6.650$ Å, an empirical estimate of the resistivity ρ_{MI} at the metal-insulator transition is obtained using Ansatz 4.1. In mnemonic notation, the resistivity ρ_{MI} at the metal-insulator transition is thereby given by the following definition:

Definition 4.2 $\rho_{MI} \equiv 258\, (\lambda_{MI}/\text{Å})\, (k_F\, l_e)_{MI}^{-1}\ \mu\Omega\ \text{cm} \approx 981\ \mu\Omega\ \text{cm}$

The fractional number 7/4 for $(k_F\, l_e)_{MI}$ at the metal-insulator transition is only approximately determined by experiment. Also, the number does not necessarily have to be written as a ratio of two natural numbers. Admittedly, this view is influenced by the fractional quantum Hall effect, where a two-dimensional electron gas generated in a semiconductor heterostructure behaves as if the charge carriers had only fractions of the elementary charge. Exactly quantized fractional Hall steps have been an experimental fact since 1982 at high magnetic field strengths (≈ 20 Tesla) and low temperatures (≈ 1 K) and are no theoretical speculations. The fractional quantum Hall state is an incompressible quantum fluid with strongly interacting charge carriers. Why should topological properties (invariants) of quantum liquids not be noticeable at the metal-insulator transition? Although quantum physics sometimes seems magical, it obeys basically very simple rules. The fractional number 7/4 will play an important role in later chapters. For the time being, it is nothing more than an arbitrarily determined ratio that "matches" experimental data.

Further experimental evidence for the critical resistance ρ_{MI} is rare except for metallic liquids. Optical conductivity measurements on "bad metals" or high-temperature superconductors, which show no or only low saturation of resistance at high temperatures, show that at a critical temperature the optical conductivity at frequency zero becomes very flat. At the transition a critical value of about 1,000 $(\Omega\ \text{cm})^{-1}$ is attained [4.3].

The length λ_{MI} of ≈ 6.650 Å is ubiquitous, since it can only be calculated by natural constants and probably marks the transition to short-range order. Under this length the matter is ordered and the deviation from a regular structure is small. Above, it is no longer possible to speak of a well-formed lattice over large areas comparable to crystals, but networks of ordered blocks are formed that develop collective behavior. Liquids, molten metals or glasses are examples of this. It therefore makes sense for a broader experimental confirmation of λ_{MI} to take a closer look at the behavior of glasses.

4.1 Properties of glasses

From x-ray diffraction (XRD) patterns of quartz glass, it is possible to deduce crystallite sizes of ≈ 7 Å by means of the width of the diffraction lines [4.4]. This corresponds approximately to the size of the unit cell of crystalline quartz. Since both scales are comparable, the crystallite size in quartz glass is reduced to the size of the unit cell of crystalline quartz. It is therefore not very meaningful to conceive quartz glass as a polycrystalline material, but it is more appropriate to view this material as a network of ordered blocks of crystallites of size ≈ 7 Å. At least for quartz glass it is well confirmed that the size of an ordered block corresponds approximately to the universal length λ_{MI}.

At high temperatures, the mean free path of sound quanta in glasses must be on the order of the disorder of the structure of the material [4.4]. By analogy with the corresponding expression in the kinetic theory of gases, the thermal conductivity κ in a strongly disordered solid can be expressed by the following formula:

Formula 4.2 $\kappa = (1/3)\, C \langle v_{sound} \rangle\, l_{Ph}$

With experimental values for κ, the heat capacity per unit volume C and the average velocity of the sound $\langle v_{sound} \rangle$, Formula 4.2 can be used to calculate the mean free path l_{Ph} of phonons. At room temperature and at 100 °C, the value for quartz glass is ≈ 6 Å [4.5]. For crown glass and flint glass, about 40% smaller values are obtained for the phonon mean free path l_{Ph} [4.4]. Is it not remarkable that a thermodynamic view of the mean free path of sound quanta requires a scale on the order of λ_{MI} in quartz glass?

4.2 The Lorenz number for metals

The empirical rule of Wiedemann and Franz states that for most metals the ratio of thermal conductivity κ to electrical conductivity ρ_{el}^{-1}, visualized by Formula 4.3

Formula 4.3 $\kappa\, \rho_{el} \sim T$

is directly proportional to temperature. The proportionality factor is called Lorenz number L_{Lorenz}. If only charges and no phonons are responsible for the heat transport, that is, electrons transport both heat and charge, the Lorenz number is almost material-independent for most metals. With the help of the model of the free electron gas, the theoretical expression $\pi^2\, 3^{-1}\, (k_B/q_e)^2$ for L_{Lorenz} can be derived yielding a universal value of $\approx 2.44 \cdot 10^{-8}$ W Ω K^{-2}. The derivation which is also called Drude-Sommerfeld theory assumes the mean free path for heat and charge transport to be equal. The Wiedemann–Franz proportionality is experimentally confirmed for many metals over a large, not at too low temperature range. At 273(373) Kelvin the experimental Lorenz numbers [4.6] in units of 10^{-8} W Ω K^{-2} are 2.31(2.37) for silver, 2.23(2.33) for copper, 2.35(2.40) for gold and 2.31(2.33) for zinc.

Although the Lorenz numbers of the Drude–Sommerfeld theory are for many materials close to the experimental values, for the best electrical conductors such as silver and copper they are clearly too high. The scatter of the measured values may lie in the fact that the heat conduction by phonons is not negligible or the measurement of the thermal conductivity κ is too inaccurate. Because of large departures from the Wiedemann–Franz law, which is a simple proportional relationship, Smith and Palmer proposed a linear relationship of the form "$\kappa = L_{Lorenz}\, (T/\rho_{el}) + const$" for fitting measurements of different materials to theory (Smith–Palmer equation). For liquid metals and molten alloys, R. W. Powell [4.7] showed that the measurements conformed very well to a straight line, and L_{Lorenz} obtained values of about the theoretical order. He determined a Lorenz number of *2.32·10⁻⁸ W Ω K⁻²*, which is close to the theoretical value of the Drude–Sommerfeld model indeed.

If the term $(7/4\; k_B/q_e)^2$ for the Lorenz number L_{Lorenz} is chosen, a value of $\approx 2.27 \cdot 10^{-8}$ *W Ω* K^{-2} is calculated being for most metals closer to experimental observations than the Drude–Sommerfeld value. The Lorenz number $(7/4\; k_B/q_e)^2$ approximates the value obtained from the Smith–Palmer equation for liquid metals equally very well. Is this a coincidence or a confirmation of the physical relevance of the fractional number 7/4 that already appeared for $(k_F\, l_e)_{MI}$ at the metal-insulator transition?

4.3 The rule of Pictet and Trouton

At the liquid/gaseous phase transition, the order and, therefore, the entropy of the system change. According to the experimentally established rule of Pictet and Trouton, the molar entropy of vaporization $(\Delta S)_{vap}$ of most nonpolar, non-associating liquids is approximately equal to ≈ 88 J/K/mol at the boiling point under standard conditions [4.8]. The Pictet–Trouton rule also allows a simple estimation of the entropy of vaporization of pure metals. For example, at atmospheric pressure, the experimental entropy of vaporization $(\Delta S)_{vap}$ of lithium (Li) or aluminum (Al) is ≈ 91 J/K/mol [4.9]. At the transition from liquid to vapor, there is a similar increase in disorder for many, but not necessarily all substances. If the empirically based value of ≈ 88 J/K/mol is normalized with the universal gas constant $R_{gas} = k_B N_{Avogadro}$ (≈ 8.314 J/K/mol), a value of ≈ 10.6 or ≈ 3 (7/2) is obtained.

Is it not surprising that the transition from liquid to vapor involves the fractional number 7/2 per degree of freedom, whereas the metal-insulator transition requires the fractional number 7/4? The cause of the empirical rule of Pictet and Trouton is unknown, why the entropy of vaporization of many liquids increases by the same amount during evaporation. The rule has remained until today what it has always been: an experimental fact without a microscopic theory.

4.4 The Mooij rule

The empirically established Mooij rule [4.10] is a criterion that relates the sign of the temperature coefficient of resistivity (TCR) with the value of the resistivity itself. By comparing many measurements of disordered transition metal alloys, J. H. Mooij found that resistivity and TCR are correlated. The rule states that at a specific electrical resistance (ρ_{Mooij}) between ≈ 100 μΩ cm and ≈ 150 μΩ cm a sign change of the TCR occurs. That is, many material compositions have a low TCR @ RT close to zero only if the resistivity is 100–150 μΩ cm. A thin layer, for instance, with a thickness of 25 nm and a resistivity of 150 μΩ cm has a sheet resistance of 60 Ω/sq. Sheet resistances of this magnitude play a central role in thin film technology. One of the best known representatives of metal alloys that have a low TCR is the metal alloy Ni_xCr_{1-x}. Not only Ni_xCr_{1-x} shows a low temperature dependence of resistance, but other metal alloys as well. The Mooij rule is valid for both thin and massive layers. This implies that the dimensionality does not play an important role. The validity of the rule is often limited to a narrow temperature range, and in the vicinity of the TCR switching, the resistivity is usually only slightly temperature-dependent.

The empirical correlation between the specific resistance and the TCR observed by J. H. Mooij provoked a controversy in the literature [4.11] as to whether this can be taken as a universal behavior. Is it always the case for disordered alloys that for a small TCR around room temperature a resistivity of approximately 150 μΩ cm is mandatory? Many observational data suggest that this statement is correct. But what is the reason for it? Does the observation, that the Mooij rule is only weakly dependent on material composition, imply universality? The physical cause of the rule, although often theoretically questioned, still remains unexplained. Using, in Ansatz 4.1, the Bohr radius a_H of ≈ 0.529 Å for λ_F, and 1 for the dimensionless parameter ($k_F l_e$), the specific resistance ρ_{Mooij} is given by the following:

Ansatz 4.2 $\rho_{Mooij} = 258\ (a_H/\text{Å})\ \mu\Omega\ \text{cm} \approx 136\ \mu\Omega\ \text{cm}$

Is it not surprising that the use of the universal length a_H along with the Ioffe–Regel criterion, results in a resistivity by means of Ansatz 4.2, that is close to the Mooij criterion?

The Evanohm® precision resistance alloy is a nickel (75%)-chromium (20%)-aluminum (2.5%)-copper (2.5%) compound that has a very low TCR (≈ 1 ppm/K) in the range from −60 °C to 120 °C. In this material, the electrons and the phonons appear to be decoupled over a very wide temperature range because of the weak temperature dependence. According to the data sheet, the electrical resistivity of the Evanohm® alloy is ≈ 134 μΩ cm [4.12]. Is it a coincidence that the Evanohm® resistivity calculated by Ansatz 4.2 pretty well matches the value of the Mooij criterion? Is this also a confirmation of the correctness of the phenomenological Ansatz 4.1 or are both agreements mere coincidences?

The two universal lengths a_H and λ_{MI}, which depend only on natural constants, are linked by an identity. It is easy to prove that

Relation 4.2 $$4\pi\, a_H = \lambda_{MI}$$

holds.

4.5 Amorphous carbon

Kazuyuki Takai and coworkers [4.13] investigated the influence of heat treatments on the electronic and structural properties of amorphous carbon thin films. The carbon was deposited on quartz glass substrates by means of a pulsed Nd: YAG laser from a graphite target in a high vacuum chamber. For heat treatment and characterization, approximately 1 μm thin films thus prepared were removed from the quartz glass substrates with a razor blade and organic solvents. The samples were then baked in vacuum under low pressure at temperatures of 200 °C, 400 °C, 600 °C, 800 °C, 1,100 °C, 1,300 °C and 1,500 °C, respectively. The heat treatment causes a gradual conversion of sp^3 to sp^2 hybridized carbon, and between 800 and 1,100 °C a continuous metal-insulator transition is observed. At 1,500 °C, sp^2 hybridization is almost complete, and the sample is likely to contain few sp^3 hybridized carbon defects. This is evidenced by the fact that the basal plane distance obtained from XRD data decreases continuously with increasing temperature and reaches the value of graphite of ≈ 3.347 Å at 1,500 °C. The resistivity of such a graphitic disordered carbon system is ≈ **10^3 μΩ cm**, and the temperature dependence between 4 K and 300 K is vanishingly small.

Isn't it amazing that a system consisting of a three-dimensional network of randomly distributed graphitic sp^2 domains has a resistivity close to the resistivity of the metal-insulator transition given by Definition 4.2? It seems that disordered graphitic sp^2 domains thermodynamically stabilized by heat treatments generate an electronic coherence analogous to metallic liquids.

An astonishing result arises when the lattice plane distance of graphite is compared with the Bohr radius of the hydrogen atom a_H of ≈ 0.529 Å. For example, for the basal plane distance in units of a_H, approximately 1.007·2π is determined. Does the fundamental quantity $2\pi a_H$ represent the minimal expansion in the z direction[3] that is physically necessary to confine charge carriers in two dimensions? Interestingly, in a bundle of single-wall carbon nanotubes (SWNT), the individual tubes have the same distance of ≈ 3.4 Å. Likewise, the shells of a multi-wall carbon nanotube (MWNT) have the same distance of ≈ 3.4 Å. The tubes in a bundle of SWNT and the shells of a MWNT seem to be stabilized by the same interaction forces as the planes in graphite. In all cases, the distance is $\approx 2\pi a_H = \lambda_{MI}/2$.

References

[4.1] Yoseph Imry. Introduction to mesoscopic physics, 2nd edn. Oxford University Press, New York, p. 9

[4.2] A. F. Ioffe and A. R. Regel. Non-crystalline, amorphous and liquid electronic semiconductors. Progress in Semiconductors 4 (1960) 237–291

[4.3] N. E. Hussey, K. Takenaka, and H. Takagi. Universality of the Mott-Ioffe-Regel limit in metals. arXiv:cond-mat/0404263

[4.4] Charles Kittel. Interpretation of the thermal conductivity of glasses. Physical Review 75 (1949) 972–974

[4.5] A. C. Anderson. Thermal conductivity. Topics in Current Physics: Amorphous Solids: Low-Temperature Properties. Edited by W. A. Phillips, Springer Verlag Berlin 1981

[4.6] C. Kittel. Introduction to solid state Physics. 5th edn. Wiley & Sons New York (1976)

[4.7] R. W. Powell. Correlation of metallic thermal and electrical conductivities for both solid and liquid phases. International Journal of Heat and Mass Transfer 8 (1965) 1033–1045

[4.8] Bimalendu N. Roy. Fundamentals of classical and statistical thermodynamics. John Wiley & Sons Limited 2002

[4.9] Erhard Hornbogen und Hans Warlimont. Struktur und Eigenschaften der Metalle und Legierungen. Springer Verlag Berlin

[4.10] J. H. Mooij. Electrical conduction in concentrated disordered transition metal alloys. Physica Status Solidi (a) 17 (1973) 521–530

[4.11] C. C. Tsuei. Nonuniversality of the Mooij correlation – the temperature coefficient of electrical resistivity of disordered metals. Physical Review Letters 57 (1986) 1943–1946

[4.12] MatWeb Material Property Data. http://www.matweb.com, queried on October 5, 2011

[4.13] Kazuyuki Takai, Meigo Oga, Hirohiko Sato, Toshiaki Enoki et al. Structure and electronic properties of a nongraphitic disordered carbon system and its heat-treatment effects. Physical Review B 67 (2003) 214202, 11 pages

3 Two-dimensional electron gases formed at the interface of hetero-structures have an extension of approximately 100 Å in the z direction. Such systems reduce the number of electrons in the z direction much less than graphite with only one carbon atom in this direction.

5 Calculation of universal parameters by means of the David Bohm analogy

Science should produce ideas and hypotheses that can be proven or refuted by experimentation. (*Karl R. Popper*)

5.1 The universal London length

The London brothers (Fritz and Heinz London) were the first to quantitatively study the phenomenological fact that a superconducting metal does not allow a magnetic field to penetrate into it [5.1]. In a superconductor, the magnetic field penetrates only slightly into the material and decays exponentially within the material. The London brothers phenomenologically introduced the penetration depth λ_{London} on the basis of experimental evidence. This characteristic measure is explained as follows:

Formula 5.1 $\lambda_{London}^2 = \varepsilon_0\, c^2\, q^{*-2}\, m^*\, n^{*-1}$

They thus succeeded in understanding essential aspects of superconductors in magnetic fields macroscopically, including the Meissner effect. They also hypothesized that superconductivity is a quantum mechanical effect that can manifest itself unexpectedly in macroscopic areas as a collective phenomenon (emergence). The theory says nothing about the mechanism of superconducting behavior, and links an individual particle with the macroscopic particle number density n^* of charge carriers. The individuality of the particles is completely lost. Below the critical temperature, typical experimental values for the London penetration depth λ_{London} are between 20 and 100 nm. Experimentally, the penetration depth can be measured by examining thin superconducting disks.

The group $\varepsilon_0\, c^2\, q^{*-2}\, m^*$ in Formula 5.1 has the dimension of a reciprocal length. Substituting $m^* = \alpha^2\, m_e$ and $q^* = q_e$ into Formula 5.1, the group $\varepsilon_0\, c^2\, q^{*-2}\, m^*$ corresponds to the already known reciprocal length λ_{MI}^{-1} set by Definition 4.1. If $n^* = n$ is also used in Formula 5.1, then penetration depth λ_{Lo} according to

Definition 5.1 $\lambda_{Lo}^2 \equiv 2^{-1}\, c\, h^{-1}\, \alpha\, m_e\, n^{-1} = (\lambda_{MI}\, n)^{-1}$

follows. The choice of the length scale λ_{MI} and the regrouping of Formula 5.1 to Definition 5.1 may seem arbitrary. However, the length scale λ_{MI} is universal, that is, material independent, and, as explained in Chapter 4, important in describing the resistance behavior of graphitic-disordered carbon and metallic liquids. In glasses, the length scale λ_{MI} seems to play an important role as well. The atomic energy scale ε_{atomic}, given by $2^{-1}\, \alpha^2\, m_e\, c^2$ ($\approx 2.18 \cdot 10^{-18}$ J), agrees with the ionization

https://doi.org/10.1515/9783110612387-005

energy (≈13.6 eV) of the hydrogen atom for infinite proton mass and is the scale at which the physics of atoms takes place. In addition, Definition 5.1 includes only geometric quantities. Further, as shown in Chapter 8, the mathematical structure visualized by

Relation 5.1 $\lambda_{Lo}{}^2 = (\lambda_{MI}\, n)^{-1} \leftrightarrow \lambda_{MI} = (\lambda_{Lo}{}^2\, n)^{-1}$

will play an important role for other lengths as well. Also, no new parameters are introduced. All these are reasons as to why the London length λ_{Lo} is defined according to Definition 5.1.

5.2 The elementary volume and the calculation of the reference temperature

First, relations that are needed for the calculation of the reference temperature τ are compiled from previous chapters. It follows from Chapter 2 that

$$\Lambda = h\, c\, 2^{-1}\, \pi^{-1/3}\, E_\tau{}^{-1}$$

$$a_B = \alpha\, (4\pi)^{-1}\, \Lambda$$

$$m_o = h\, c^{-1}\, \Lambda^{-1}$$

and from Section 3.2

$$n^{1/3}\, a_B = c_1 = (4\pi^{4/3})^{-1/3}$$

as well as from Chapter 4 and Section 5.1

$$\lambda_{Lo}{}^2 = 2^{-1}\, c\, h^{-1}\, \alpha\, m_e n^{-1} = (\lambda_{MI}\, n)^{-1}$$

These relations have the same unit on both sides of the equal sign and express a correct physical fact. They are formed from known natural constants and the six quantities Λ, E_τ, a_B, m_o, n and λ_{Lo}. With Hypothesis 1.1, the reference temperature τ could be calculated by means of these relations if a further reference length were available for the calculation of the reference volume V. The missing length, called L, shall heuristically be defined by

Definition 5.2 $L \equiv 4^{-1}\, \alpha^{-1/2}\, \Lambda$

As a vague justification for the choice of the quantity $\alpha^{1/2}$, it should be mentioned that $\alpha^{1/2}$ plays a similar role in the formulas of quantum field theory as the Newton constant G and c in the formulas of general relativity. However, the fact that the determination of the reference length L by Definition 5.2 also makes sense physically is only apparent in later chapters. With Definition 5.2 an elementary cuboid with the volume $\Lambda\, L\, \lambda_{Lo}$ can now be formed and Hypothesis 1.1 becomes

Equation 5.1 $m = m_0 = h\, c^{-1}\, \Lambda^{-1} = V/(8\pi\, c^2) = (\Lambda\, L\, \lambda_{Lo})/(8\pi\, c^2)$

which can be used for the determination of the reference energy E_τ or the reference temperature τ that defines the statistically distributed heat energy in the volume $\Lambda\, L\, \lambda_{Lo}$. By transforming Equation 5.1 and assuming that the kilogram can be equated to $m{\cdot}s^2$, the reference energy E_τ is, after some calculation, given by

Relation 5.2 $E_\tau = 2^{-8/3}\, c^{8/9}\, h^{2/3}\, \alpha^{1/3}\, \pi^{-20/27}\, m_e^{1/9} \approx 1.56{\cdot}10^{-20}\, m^3$

Due to the radical assumption that $J = kg\, m^2\, s^{-2} = m^3$ holds, Relation 5.2 is equivalent to $E_\tau \approx 1.56{\cdot}10^{-20}$ J or

Relation 5.3 $\tau = E_\tau/k_B \approx 1{,}133.93$ K

With Relation 5.2 or 5.3, all the relevant quantities that depend on E_τ can now be calculated only from natural constants. They have the values given in Table 5.1 and arise out of the radical assumption of Hypothesis 1.1. Is E_τ the "hidden" heat energy or the heat bath of the vacuum proposed by L. De Broglie [5.2], which maintains itself and to which all quantum particles are coupled? Unfortunately, the idea of L. De Broglie plays no role in current doctrine and the existence of a Lorentz-invariant heat medium is completely ignored.

In Relation 5.2, the reference energy E_τ is represented as the product of powers with rational exponents of the Planck constant h, the speed of light in vacuum c, the mass m_e of the electron, the Sommerfeld constant α and the mathematical constants 2 and π. Such a description is basically feasible for other quantities as well. They all have in common that they can be represented as rational powers of the quantities h, c, m_e, k_B, the Sommerfeld constant α and the number constants 2 and π. In addition to τ, the power products of the length scale Λ and the particle number density n are also listed in Table 5.1.

The values of the fundamental lengths, tabulated in Table 5.1 and determined by the reference temperature τ, have scales that are experimentally accessible and are involved in the structuring of matter. It is striking that the two lengths L and Λ belong to the micrometer scale, which is essential in magnetic domains. This length scale is also important for colloid particles. Polymer molecules also reach the micrometer scale at high molecular weights.

What lies behind the values that have been gained by the described formalism? Do these variables have an invariant character? Is it possible to conclude from these values

Table 5.1: Fundamental quantities.

Qty	Value	Unit	Power products	Geometrized unit
T	≈1,134	K	$c^{8/9}\,h^{2/3}\,m_e^{1/9}\,\alpha^{1/3}\,2^{-8/3}\,\pi^{-20/27}\,k_B^{-1}$	$[c^{8/9}\,h^{2/3}\,m_e^{1/9}] = m^3 = J$
Λ	***≈4.33***	µm	$c^{1/9}\,h^{1/3}\,m_e^{-1/9}\,\alpha^{-1/3}\,2^{5/3}\,\pi^{11/27}$	$[c^{1/9}\,h^{1/3}\,m_e^{-1/9}] = m$
a_B	≈2.52	nm		
m_0	$\approx 5.1\cdot10^{-37}$	kg		
n	$\approx 3.41\cdot10^{18}$	cm^{-3}	$c^{-1/3}\,h^{-1}\,m_e^{1/3}\,\alpha^{-2}\,2^{-1}\,\pi^{4/9}$	$[c^{-1/3}\,h^{-1}\,m_e^{1/3}] = m^{-3}$
λ_{Lo}	***≈21***	nm		
L	***≈12.7***	µm		

Note: The numerical values of the dimensions of the elementary volume of constant heat energy are in italics and bold accentuated. For the units of the power products, the unit kg must be equated consistently with the unit m s^2.

on properties of matter? Do these "conserved quantities" have a fundamental meaning at all? Can they be used for the explanation of structure formation? Does the axiomatic thermodynamics, which describes systems by means of energy, volume and temperature, and determines how much energy is available, play a much larger role for the physical world than is suspected? Do we underestimate the importance of thermodynamics, which deals with transformations of energy into different forms, for the physical world?

The elementary volume $\Lambda\, L\, \lambda_{Lo}$ has a value of about 1.2 µm³. Does this value provide information about where the quantum world of the small ends and the macroworld begins with the classical physical laws? In particle physics, thermodynamics with its three axiomatic laws does not play a major role, since the calculation of the Lagrange density is based solely on the potential and kinetic energies, and the thermodynamic energy or thermal energy is completely excluded. The formalism of particle physics is exclusively a dynamic theory that does not allow quantitative predictions of its parameters.

5.2.1 The best electrical conductor silver

Silver has the lowest resistivity of all materials. Its value [5.3] is about 1.59 µΩ cm at T = 20 °C. Although the silver electrons can move relatively freely due to the small resistivity, no superconducting macrostate, as with all alkali and precious metals, has ever been found. Only poorly conductive metals such as lead or mercury form this exceptional quantum state. Good metals such as silver must have physical properties that are not favorable for superconductivity compared to bad metals. In the silver atom a single electron "moves" in the Coulomb field of 47 protons, which are screened by 46 electrons. With dimensional considerations and the experimental fact that the charge transport in a metal occurs by electrons

Ansatz 5.1 $\rho_{silver} = (h/q_e^2)\,(k_B T)\,(m_e c^2)^{-1}\,L$

can be deduced. Ansatz 5.1 contains the resistance quantum h/q_e^2, the characteristic length L and the ratio of the measuring temperature T to the rest energy of the electron. It also incorporates the linearity of the temperature behavior, which is proven for silver for temperatures greater than the Debye temperature, and is observed for other metals as well. By using CODATA values for natural constants, Ansatz 5.1 yields for T = 293 K about **1.62 µΩ cm**, which in fact corresponds approximately to the specific resistance of silver of ≈**1.59 µΩ cm** at 20 °C. Is it not remarkable that the resistivity of silver can be calculated with respectable accuracy only by constants of nature, the fundamental length L, and under the assumption that resistivity is a linear function of temperature? Ansatz 5.1 not only distinguishes the characteristic length L as a fundamental quantity, but also indicates that in total three quantities have great physical relevance in the description of resistivity: temperature, mass and length.

5.2.2 Bad metals

A class of new materials in solid-state physics, known as bad metals, show puzzling transport properties that cannot be understood with the well-known transport theories. For example, the electrical resistance continues to increase linearly even at high temperatures without saturating as is common in classical metals. Also, the frequency-dependent optical conductivity, which can be measured by infrared or terahertz time-domain spectroscopy, has a fundamentally different appearance than that predicted by popular beliefs. Such materials are of great scientific and technical interest and their understanding is one of the central problems of today's condensed matter physics. They show phase transitions in states with different charge and spin configurations or in the superconducting state, which occurs especially in bad metals and not, as intuitively suspected, in materials with the best electrical conductivities.

S. B. Arnason and coworkers [5.4] mimicked the behavior of bad metals on glass with the good metal silver (Ag). Depending on the microstructure, thin films of silver have extremely high resistivities with anomalously large temperature dependencies and oddly enough negative magnetoresistances, which are quadratic in the applied magnetic field. These are properties that are associated with materials referred to as bad metals. The films were grown in vacuum by thermal evaporation of silver onto glass substrates at room temperature. During vapor deposition, the sheet resistance was measured in situ as a function of the thickness, which was determined with a quartz crystal monitor calibrated by atomic force microscopy. The growth process can thus be monitored as a function of the amount of material deposited. Initially, the layer consisted of isolated silver clusters and finally formed a homogeneous film of microscopic grains with increasing thickness. During the growth process, a phase transition as a function of coverage was visible, which could be identified by a distinctive change in resistance behavior during the transition. By a power law scaling of the resistance, S. B. Arnason and coworkers determined the transition at a critical film thickness of ≈19.2 nm. Further observations and explanations of the two authors on this subject can be found in [5.5]. Can the transition observed by S. B. Arnason and colleagues be attributed to the length λ_{Lo} (≈21 nm), that is, does λ_{Lo} establish the relevant length scale of the experiment?

5.2.3 Metal clusters

A fundamental question of solid-state physics is how many atoms are necessary for an atomic assembly to behave macroscopically like an insulator, a metal or a superconductor, that is, to show solid-state properties. With the aid of a metal cluster source, crystalline nanoparticles of well-defined size ($L_{cluster}$) can be condensed together with an inert gas (Kr) as a matrix onto a cold substrate [5.6]. The granular films thus formed can be examined by various physical methods. Weitzel and coworkers [5.7] found that small crystalline Bi clusters are superconducting, whereas the bulk material with the same rhombohedral structure is not superconducting. They determined the superconducting transition temperature T_c as a function of the mean cluster size (2.5 nm < $L_{cluster}$ < 40 nm), at which the local accumulations became superconducting. The transition temperature T_c is clearly visible experimentally and the cluster size can be accurately determined with the help of transmission electron microscopy (TEM). The superconductivity started with $T_c \approx 5.5$ K for Bi clusters with $L_{cluster} \approx 2.5$ nm and stopped with $T_c \approx 2$ K for $L_{cluster} \approx 20$ nm. The superconductivity does not disappear after annealing of the Bi cluster at 300 K and cooling again to low temperatures. The cluster-size-dependent normal-state resistivity has a pronounced maximum at ≈5 nm. All investigated films were far above the percolation threshold with a metal volume fraction of ≈0.9. The typical film thicknesses were ≈150 nm. Is the cluster-size-dependent superconductivity observed by Weitzel and coworkers induced by the length λ_{Lo} (≈21 nm)?

5.3 Manifestations of the reference temperature in the physical reality: a selection

The following examples are intended to show that the reference temperature τ given by Relation 5.3 could actually play a role as an experimental reference scale. They are a selection found by literature search. Not all examples are universal in nature and of course not free from a subjective point of view. The existence of the fractional number 7/4 or its reciprocal is also supported by further examples.

5.3.1 The critical temperature of water

There is a thermodynamic state where the difference between the gas and the liquid ceases to exist because the mean distance between the particles is comparable in both phases. This critical point is characterized by a critical temperature, a critical pressure and a critical density. The state variables at the critical state are characteristic constants for each substance and can be determined approximately using the Van der Waals constant and the gas constant R_{gas}. The critical temperature of water [5.3] is ≈374.1 °C or ≈4/7 in units of the reference temperature τ.

5.3.2 The Debye temperature of diamond

The experimentally accessible Debye temperature Θ_D is a characteristic quantity for describing the temperature dependence of the heat capacity of a material. It can also be used as a measure of the hardness and is linked to the largest angular frequency in the spectrum of sound waves that can propagate in an elastic medium with continuous mass distribution. The Debye temperature, which can also be regarded as a limit temperature, allows a universal representation of the temperature dependence of the specific heat with only one material-specific parameter. The experimental Debye temperatures cover a range from 23 K at a pressure of 25 bar (21 cm^3/mol) for solid helium [5.8] to 2,250 K for diamond [5.3]. Solid helium is the softest, and diamond is the hardest known crystalline solid. Remarkably, the Debye temperature of diamond at low temperatures is about twice as high as the temperature τ.

5.3.3 The phonon density of states of polycrystalline diamond

With high-intensity synchrotron sources, it is possible to measure phonon densities of states of ordered/disordered solids or liquids using crystal optics. By measuring inelastically scattered x-ray radiation on phonons, an insight into the dynamics of such systems is also obtained. The vibrational density of the states of polycrystalline diamond[1] measured [5.9] by inelastic x-ray scattering drops to zero at a cutoff energy ε_{cut} of ≈170 meV. If the cutoff energy in units of eV is estimated by

Ansatz 5.2 $$\varepsilon_{cut} \equiv 7/4\,(k_B/q_e)\,\tau$$

1 The average grain size of the measured sample was 3–5 μm, and the effective scattering volume amounted to 0.08 mm^3.

a value of ≈171 meV results, which is equivalent to ≈41.35 THz and agrees very well with experiment. It is worth noting that also the fractional number 7/4 crops up in Ansatz 5.2. The longitudinal velocity of sound in diamond [5.9] is ≈18,240 m s^{-1}, which corresponds to the highest velocity of sound observed in all materials. The relationship $l_{ph} = v_{sound}\,(2\pi\, f_{cut})^{-1}$, using f_{cut} ≈41.35 THz, gives a coherence length l_{ph} of ≈0.702 Å or ≈4/3 in units of a_H. Surprisingly, the double of the coherence length $2 \cdot l_{ph}$ (≈1.41 Å) corresponds roughly to the shortest interatomic distance (≈1.42 Å) of carbon in a single crystal of graphite in the planes. The shortest interatomic distance of carbon in diamond is ≈1.54 Å.

5.3.4 The Raman G line of graphite

In Raman spectroscopy, a sample is irradiated with monochromatic laser light and the inelastically scattered light is analyzed by means of a spectrometer. In such an experiment the photon impinging on matter loses or gains energy from virtual states and is simultaneously scattered. In contrast to the elastic Rayleigh scattering, the wavelength of the scattered photon shifts. If the scattered light has a wave number smaller than that of the incident light, the lines in the spectrum are called Stokes lines. If the scattered light has a larger wave number than that of the incident light, the lines are called anti-Stokes lines. A wave number is defined as the reciprocal of the wavelength in centimeters and is thus directly related to energy. Classically, the process is based on a change in the polarizability of the material by the incident light wave. The mass of the atoms, their mutual bonds and the periodicity of the lattice play a crucial role in Raman spectroscopy.

F. Tuinstra and coworkers [5.10] showed that the Raman spectrum of a large, single crystal of natural graphite exhibits a sharp, intense peak ($\varepsilon_{graphite\_G\text{-}band}$) at ≈1,575 cm^{-1}. Because of the focusing and absorption of the used argon ion laser light (488 nm), the observed Raman spectrum was induced in an illuminated circular area with a diameter of ≈20 µm and a depth of ≈50 nm. On this grounds, the single crystals need to be larger than the area illuminated, that is, ≈20 um. F. Tuinstra and coworkers could not prove a dependence of the line on the basal plane distance, that is, the interlayer distance between consecutive planes. Also different orientations of the crystal with respect to the incident beam always gave the same narrow single line without polarization effects. Smaller crystals slightly shifted the line toward higher wave numbers.

R. J. Nemanich and S. A. Solin [5.11] observed the line at ≈1,581 cm^{-1} in samples of artificially grown, highly ordered pyrolytic graphite (HOPG).[2] Yan Wang and

2 HOPG is an artificially grown graphite with an almost perfect alignment perpendicular to the carbon planes. Along the in-plane directions the crystallites are small and randomly oriented. A highly oriented pyrolytic graphite differs, therefore, significantly from a single crystal of graphite.

coworkers [5.12] studied several graphite materials and reported the position of the line at ≈1,577 cm^{-1} in HOPG when the samples were excited at 458 nm. When excited at 515 nm, they observed the line at ≈1,580 cm^{-1}. Recent studies [5.13] on free-standing monolayers of graphene,[3] suspended over micrometer-sized trenches, revealed a line at ≈1,577 cm^{-1} at low excitation powers. Free-standing samples provide an experimental environment detached from doping induced by the substrate or other influences.

If the wave number 1,577 cm^{-1}, at which the pronounced transition of the G band in graphite takes place, is converted into a vibration temperature by multiplication with hc/k_B, a value of ≈2.001 in units of τ results. Even if the measurement of the position of the G line is subject to experimental fluctuations and environmental influences, this reveals an impressive connection to the reference temperature τ given by Relation 5.3. Is it a coincidence that the Debye temperature of diamond is also about twice as large as τ?

5.3.5 The Raman line of silicon

The tetravalent semiconductor silicon (Si) crystallizes like carbon (C) in a cubic-face-centered lattice. Due to strong covalent bonds, there are no free electrons available due to the saturated valences. Defect-free Si and C are therefore ideal insulators at temperatures close to zero. T. R. Hart and coworkers [5.14] studied the Raman spectrum of single crystals of silicon in the temperature range of 20–770 K and observed that the Stokes line, using the 514 nm Ar laser line for excitation, approached a limiting wave number of ≈525 cm^{-1} at low temperatures. A special feature of the Raman spectrum of Si is the strong dependence of the intensity of the anti-Stokes line on temperature. At high temperatures, Stokes and anti-Stokes lines are comparable in intensity. At low temperatures, the anti-Stokes line is so weak that it can hardly be detected.

Using silicon nanowires immersed in suprafluid helium (≈2 K), H. Scheel and coworkers [5.15] observed the Stokes line at ≈523 cm^{-1} when the samples were illuminated by moderate laser powers at 514 nm. If the mean value (524 cm^{-1}) of the experiments mentioned above is converted into a vibration temperature, ≈0.665 or ≈2/3 results in units of τ.

5.3.6 The minimum thermal conductivity

The majority of nonconducting, single- or polycrystalline materials have thermal conductivities above room temperature, which have a characteristic 1/T dependence with increasing temperature. At higher temperatures, the thermal conductivities reach asymptotically constant values and behave in a similar manner to amorphous

3 Graphene consists of two-dimensional layers of carbon atoms in which charge carriers can move like particles without mass (photons) at relativistic velocities.

solids, whose thermal conductivities increase with increasing temperature and also approach a limit at high temperatures. This experimental observation supports the view that lattice vibrations of disordered crystals have to be of a similar kind as "lattice vibrations" of amorphous materials. According to an idea of Albert Einstein, the lack of a characteristic 1/T dependency is based on the fact that, in severe disorder, atomic vibrations are incoherent and the heat energy propagates in a diffusion-like manner with a mean free path approaching the mean interatomic distance. The value κ_{min} is, therefore, also called the Einstein limit.

There are various formulas for κ_{min}. All depend on experimentally determinable material parameters, such as the number of atoms per volume, and the longitudinal and the transversal velocities of sound. These formulas, however, do not provide a deeper understanding or indicate whether there may possibly exist a fundamental, universal limit for the thermal conductivity κ_{min}. The reduction of the thermal conductivity of perfect, pure solids is intuitively not possible by the mere introduction of defects or fine-grained structure.

Today, dense yttria-stabilized zirconia (YSZ) is often used as a reference for the limit of thermal transport of disordered crystalline structures. This material, whose components form a network among one another, has a thermal conductivity of about 2.5 W m^{-1} K^{-1} at high temperatures. This value has been considered exceptional for a long time and it has been suggested that there exists no dense, three-dimensional material with a lower thermal conductivity.

Recently, however, pore-free compositions of disordered crystalline systems have been found, which have thermal conductivities smaller than that of the dense YSZ ceramic. Table 5.2 lists examples of such compounds that were all prepared by sintering or hot pressing from powder mixtures.

Table 5.2: Thermal conductivities

Compound	Rel. density	κ (W m^{-1} K^{-1})	T-range T (°C)	References
7YSZ + 3.5$EuO_{1.5}$ + 3.5$TmO_{1.5}$	1	≈2	[100,1000]	[5.16]
$W_3Nb_{14}O_{44}$	1	≈1.85	[100,1000]	[5.16]
$YO_{1.5}$ + $TaO_{2.5}$ + ZrO_2	>0.97	1.7–1.9	[100,900]	[5.17]
$YbO_{1.5}$ + $TaO_{2.5}$ + ZrO_2	>0.97	1.4–1.6	[100,900]	[5.17]
$(Zr_{0.5}Hf_{0.5})_{0.87}Y_{0.13}O_{1.94}$	0.98	≈1.5	[100,700]	[5.16]
$GdPO_4$	0.98	≈1.5	[500,1000]	[5.16]
$Gd_{10}(SiO_4)_6O_3$	>0.96	1.4–1.6	[100,600]	[5.18]
$Gd_8Ca_2(SiO_4)_6O_2$	>0.99	1.4–1.6	[100,600]	[5.18]
$Gd_{8.666}Ca(SiO_4)_6O_2$	>0.99	1.4–1.6	[100,600]	[5.18]

Note: Examples of complex multicomponent compounds with very low thermal conductivities, which change only slightly within the specified temperature range. Striking is the broadness of the temperature ranges over which the thermal conductivities are nearly constant. All listed thermal conductivities are smaller than those of the dense yttria-stabilized zirconia ceramic with a thermal conductivity of about 2.5 W m^{-1} K^{-1}.

It can be seen from Table 5.2 that the lowest thermal conductivity of nonporous, polycrystalline dielectric materials at higher temperatures is on the order of magnitude of ≈1.5 W m^{-1} K^{-1}. This finding suggests that there may be a universal value κ_{min} for a high temperature limit of fully dense three-dimensional isotropic materials that could possibly be explained by

Ansatz 5.3 $\kappa_{min} = (7/4\ k_B/q_e)^2\ R_K^{-1}\ \lambda_{MI}^{-1}\ \tau \approx 1.50\ W\ m^{-1}\ K^{-1}$

which achieves a remarkable agreement with the value of ≈1.5 W m^{-1} K^{-1} extracted from Table 5.2. The value κ_{min} defined by Ansatz 5.3 depends only on natural constants and is probably the cause of the elementary energy E_τ to which the lattice is coupled. All occurring quantities in Ansatz 5.3 are universal parameters of previous chapters, such as $L_{Lorentz}$, $R_{Klitzing}$, λ_{MI} and τ.

5.3.7 The logistic model and collective magnetism

The simplest mathematical model for a dynamic nonlinear system is the discrete version of the continuous differential equation developed by the Belgian mathematician Pierre Francois Verhulst. This simple model carries the characteristic of self-similar patterns (coherent behavior) and chaotic behavior. The equation, also called "logistic model", models the natural growth and saturation of biological populations.

The solutions of the logistical growth model are dependent on a growth parameter ($0 \le g \le 4$) that describes the interplay of death and birth rates in the population and thus determines how many solutions are possible or whether chaos arises. The logistic difference equation is given by

Formula 5.2 $x_{n+1} = g\ x_n\ (1-x_n)$

where n denotes a running variable and $0 \le x_n \le 1$ represents a normalized number, which defines the population relative to the largest population that the system can sustain.

For $3 \le g < 1 + 6^{1/2}$, the solution bifurcates into two solutions. This range is also called the cycle with period 2. For $g = 1 + 6^{1/2}$ or $k = 2$, the two solutions divide into four solutions that double with each additional step of k. In Table 5.3 these particular properties are shown. Interestingly, there is a critical value for g, from which the solutions become chaotic. The increase in bifurcating becomes faster and faster in the vicinity of $g_k \approx 3.57$, and successive intervals $g_{k+1}-g_k$ become correspondingly shorter and shorter. Mitchell Feigenbaum [5.19] found that for all nonlinear quadratic transformations with a unique differentiable maximum a stable limit cycle exists and the limit ratio defined by

Formula 5.3 $(g_{k+1} - g_k) / (g_{k+2} - g_{k+1}) \rightarrow 4.669201609\ldots$ for $k \rightarrow \infty$

Table 5.3: Bifurcations of the logistic model.

k	g_k	$g_{k-1} \leq g < g_k$	τ/g_k	
			(K)	(°C)
0	1	$x_n \rightarrow 0$	1,134	
1	3	$x_n \rightarrow (1-g^{-1})$	378	105
2	$1 + 6^{1/2} \approx 3.449$	$x_n = x_{n+2}$	329	56
3	3.544 090 359 ...	$x_n = x_{n+4}$	320	
4	3.564 407 266 ...	$x_n = x_{n+8}$	318	
∞	3.569 945 672..	x_n is chaotic	318	45

Note: Bifurcation table of the logistic Formula 5.2. Also noted are the temperatures in Kelvin and degrees Celsius when the growth parameter g_k is linked to the reference temperature τ.

converges to a universal number of about 4.669. This universal number is called the Feigenbaum number today.

The metals Co, Fe and Ni and the lanthanide Gd are the only elements [5.20] that are inherently magnetized at room temperature and lose their magnetization only when the temperature rises above the ferromagnetic Curie temperature T_{Curie}. Below the ferromagnetic Curie temperature, these materials are ferromagnetic with a preferably parallel alignment of the magnetic moments, above the ferromagnetic Curie temperature, however, they are paramagnetic.

The temperature dependence of the magnetic susceptibility χ is described by the Curie–Weiss law $\chi^{-1} \sim T-\Theta_P$. The quantity Θ_P is called the paramagnetic Curie temperature. It is a measure of the interaction energy between elementary magnetic dipoles and is always greater than the ferromagnetic Curie temperature T_{Curie}. In Table 5.4 the temperatures T_{Curie} and Θ_P are tabulated for all four metals.

Table 5.4: Curie temperatures.

Material	T_{Curie} (K)	Θ_P (K)	τ/Θ_P	Remarks
Co	1,396	1,415	≈ 0.80 ≈ 4/5	
Fe	1,043	1,100	≈ 1.03 ≈ 1	
Co_2FeSi	1,100 ± 20	1,150 ± 50	≈ 0.99 ≈ 1	Highest Curie temperature in the class of Heusler compounds [5.21]
Fe_3O_4		851	≈ 1.33 ≈ 4/3	Magnetite [5.22]
Ni	631	649	≈ 1.75 ≈ 7/4	
Gd	293	317	≈ 3.58	Chaos number

Note: Curie temperatures of the ferromagnetic transition metals Co, Fe and Ni and the lanthanide Gd. As a comparison, the Curie temperature of the magnetic mineral magnetite Fe_3O_4 is shown, which also shows correlated or "entangled" behavior. In the case of magnetite, the ratio τ/Θ_P surprisingly corresponds to the ratio of the number of O atoms to the number of Fe atoms.

Comparing the paramagnetic Curie temperatures Θ_P with the reference temperature τ yields interesting dimensionless numbers, which may possibly give rise to an underlying mathematical formalism. It is astonishing that one of these numbers obtained corresponds to the chaos number of the logistic model and all the remaining represent simple fractions.

The logistic model could also play a role in the transport of charges. The conduction of charges takes place either by electrons or holes. Injected carriers are captured by localized states and released by thermal activation. Charge carriers and localized states, which can be considered as hole/electron traps, are thus in a predator-prey relationship akin to the logistic model.

5.4 The Sommerfeld constant as a number constant

> It's one of the greatest damn mysteries of physics: a magic number that comes to us with no understanding by man.
> (*Richard Feynman*)[4]

The most important of all dimensionless constants of physics is the Sommerfeld or fine-structure constant α, which cannot be measured directly in any experiment, but is always the result of other measured quantities. It decisively determines the laws of physics as they are. Today, the Sommerfeld constant is primarily the coupling constant of the electromagnetic interaction in the Standard Model of particle physics with its many parameters, all of which must be determined experimentally. The Sommerfeld constant, however, was first introduced by the atomic physicist Arnold Sommerfeld in order to calculate the spectral lines of atomic hydrogen more accurately by means of elliptical orbits. Since α is a dimensionless ratio number, the quantity has the same value in each system of units. It has almost mystical status and is still one of the greatest mysteries of physics, because it is not known how α can be theoretically calculated by deduction. This applies without exception to all dimensionless constants of physics similarly.

Dimensionally, the relation $[q_e^2/(4\pi\varepsilon_0)] = [h_{bar}c]$ applies. It postulates, according to the criterion of Albert Einstein, that a proportionality constant of order unity should exist for numerical agreement. Interestingly, in this case the proportionality constant is the Sommerfeld constant α, which should therefore theoretically be calculable by means of quantum electrodynamics (QED). But, this has not been achieved to date. What reasons does nature have for giving the Sommerfeld constant the value it has? There are many mathematical formulas that approximate the experimental value of α to many decimal places. In most cases, physical arguments for these formulas or an accepted theory for

4 QED: The strange theory of light and matter, Princeton University Press (1985), page 129.

the computation of α are missing. For a long time, however, the opinion prevails that α is of geometric origin and may possibly be explained by the use of π.

The introduction of the characteristic length L by Definition 5.2 may seem arbitrary. Also, the fact that L serves well in describing the resistivity of silver may be a coincidence. The following is intended to better anchor the length L by an attribute, which is not readily apparent. If the reference temperature τ from Relation 5.3 is geometrized by the constant hc/k_B occurring in the Planck law of radiation, and thus converted into a length

Relation 5.4 $hc/(k_B\tau) = h\,c\,E_\tau^{-1} \approx 12.688\ \mu m \approx L \approx 12.677\ \mu m$

results. Obviously, the length, which is formed by conversion of the energy $E_\tau = k_B\tau$, corresponds approximately to the length L. In other words, the energy E_τ and the reciprocal length L^{-1} are mutually defined by the two constants h and c via the famous De Broglie relation (hc/ε). If $h = c = 1$ is set, $E_\tau = L^{-1}$ naturally evolves. Because it is unlikely (unphysical) that nature uses two critical lengths close to each other, numerical (experimental) facts might require stringent equality. This can axiomatically be established by

Hypothesis 5.1 $h\,c\,L^{-1} = E_\tau = k_B\tau$

Using Definition 5.2 and Relation 2.1, this results in

Equation 5.2 $hc\,/\,(k_B\tau) = 4^{-1}\,\alpha^{-1/2}\,\Lambda = 4^{-1}\,\alpha^{-1/2}\,h\,c\,2^{-1}\,\pi^{-1/3}\,(k_B\tau)^{-1}$

from which a geometrized Sommerfeld constant α_{geom} fixed by

Definition 5.3 $\boldsymbol{\alpha_{geom}} = q_e^2/(h_{bar}\,c\,4\pi\varepsilon_0) = q_e^2\,\mu_0\,c/(2h) \equiv \mathbf{2^{-6}\,\pi^{-2/3}} \approx 1/137.28$

can mathematically be determined on the basis of Hypothesis 5.1.

Can the Sommerfeld constant be understood solely on the basis of a geometric idea (equality of lengths)? Does the simple number constant $2^{-6}\,\pi^{-2/3}$ represent the mathematical expression for this quantity that has been sought for so long? Or is Definition 5.3, like so many other proposals for the Sommerfeld constant, another α-numerology and gimmick without physical relevance and traction in the physics community? In any case, Definition 5.3 allows an independent, unambiguous comparison with experiments, since α_{geom} is not itself derived from experimental data using QED which has to be tested.

In the SI system, the limit speed of light in vacuum c and the magnetic field constant μ_0 are assigned fixed values. Thus, because of the Maxwell relationship $\varepsilon_0 \equiv 1/(\mu_0 c^2)$ and $\mu_0 \equiv 4\pi{\cdot}10^{-7}$, a fixed value is also obtained for the numerical value of the field constant ε_0. Definition 5.3 is either a conditional equation for the charge q_e as a derived quantity of h or a conditional equation for the Planck constant h as a derived quantity of q_e. Presumably, the former is physically the more sensible way, as Albert Einstein already demanded that q_e should be calculated from h and c. Another argument for the fact that q_e and not h is the dependent parameter is the evidence that in the other case the uncertainty principle of Heisenberg would have to be reconsidered, as only the Planck constant h appears in its formulation.[5]

From the definition of the Sommerfeld constant α, the dimensionless version of the electric charge given by

Formula 5.4 $\alpha^{1/2} = q_e/q_{Planck}$ with $q_{Planck}{}^2 = h_{bar}\, c\, (4\pi\varepsilon_0)$

is obtained by transformations. The quantity $\alpha_{geom}{}^{1/2}$ is interestingly associated by a simple relationship with the volume of the three-dimensional hypersphere V_3 with radius $r = c_1$ from Section 3.2. The connection between $\alpha_{geom}{}^{1/2}$ and $V_3(c_1) = 3^{-1}\pi^{-1/3}$ is given by

Relation 5.5 $8\, \alpha_{geom}{}^{1/2} = 3\, V_3(c_1) = 3\, N_{Debye}$

which can easily be verified by the use of Formula 3.6 and Definition 5.3. It is striking that the two natural numbers 8 and 3 occur, which may indicate an underlying symmetry operation, by which the volume of the three-dimensional hypersphere $V_3(c_1)$ can be converted into the dimensionless electrical charge $\alpha_{geom}{}^{1/2}$. Maybe, Relation 5.5 which represents a geometrical situation and pins charge down as a geometrical attribute can also solve the mystery, why the charge of the proton is as large as the charge of the electron and thus the hydrogen atom and the entire macrocosm are completely neutral. Somehow, the neutrality of the universe must also manifest itself in the microworld. Current theories have no explanation for why in atoms electric charges exactly cancel. It is easy to show that the identity given by

Relation 5.6 $c_1{}^3 = \alpha_{geom}{}^2\, 2^{10}$

is valid.

5.4.1 The magnetic-moment anomaly of the free electron

The current value of the Sommerfeld constant α_{Codata} is ≈ 1/137.036 and barely corresponds to the value obtained with Definition 5.3. In other words, α_{geom} is not in the

5 Because of the redefinition of the kilogram, the intention is to assign an exact value not only to the constant ε_0 and c but also to the Planck constant h. This implies that the charge q_e is also determined only by the fine-structure constant α.

tolerance range of today's value at all! This means that either Definition 5.3 is wrong or the constant α_{Codata} has currently been determined in a way that it only roughly matches the geometrized value of $2^{-6}\,\pi^{-2/3}$. It is more likely to be the former, as it is mentioned in many publications and physics books that α_{Codata} is experimentally known at least up to ten decimal places. Is that really correct? Or could it be that the Sommerfeld constant was computationally adjusted and "bent" in such a way that it only complies with the established model of QED and that every other perspective leads to a deviation? In any case, the Sommerfeld constant as a dimensionless quantity cannot depend on the choice of the fundamental constants and must be a number constant.

The Dirac equation predicts a spin g-factor of exactly 2 for the magnetic moment of the free electron as a consequence of relativistic invariance. The difference to the effectively measured value divided by 2, that is, $g_e/2–1$, is called the anomalous magnetic-moment of the electron a_e. This dimensionless number can be measured extremely accurately and the measured value a_{e_meas} is ≈0.001159652 according to CODATA. Today, the CODATA value of the Sommerfeld constant α_{Codata} is mainly derived from the measurement of the anomaly of the magnetic-moment of the electron and the underlying quantum field theory. For a pure electron-positron-photon-QED, that is, for electrons and positrons that interact with each other by means of light (photons), the theoretical result for $a_e \equiv (g_e - 2)/2$ is given by [5.23]

Formula 5.5 $$a_{e\_theo} = (1/2)\cdot(\alpha/\pi) - 0.328479...(\alpha/\pi)^2 + 1.181241...(\alpha/\pi)^3$$

up to the order of three loops. By solving Formula 5.5 for α, together with a_{e_meas}, a value $\alpha(a_e)$ for the Sommerfeld constant results, which is more or less equal to the tabulated CODATA value α_{Codata}. The value $\alpha(a_e)$ is correct only if both a precise measurement is disposable and the underlying theory (Formula 5.5) is valid. In order to be able to follow the measuring accuracy of a_e, more and more loops were included in Formula 5.5. This raises the fundamental question, especially if the agreement is shifted to ever higher decimal places, to what extent theory and experiment are still independent of each other, because the correction of the raw data, that is, the involvement of systematic errors, is also based on QED – an inadmissible circular reference because the experimenters have theoretical prejudices? The more decimal places of the experimental data agree with the theoretical calculations, the more the formalism of QED becomes an end in itself!

In addition to the determination of α by means of a_e, other methods are often given in the literature. These measurements should (must), if the physics is consistent in itself, lead to the same result, since the theories used are also based exclusively on QED. Therefore, they do not provide independent strategies for determining a model-independent value of the Sommerfeld constant because the argument runs in a circle. Perhaps it would be more appropriate to give the quantity α_{Codata} the name α_{QED}, because without QED, no high-precision value of the fine-structure constant can be determined today.

The lowest term $\alpha/(2\pi)$ in the development of a_e was first derived by Julian Schwinger. It involves a single virtual exchange of a photon with the electron and is also called one-loop correction or Schwinger correction. The deviation $(a_{e,meas} - a_{e,theo})/a_{e,meas}$ is $\approx$ 0.152 %, if only the one-loop correction $\alpha_{Codata}/(2\pi)$ is taken into account. Although the Schwinger term $\alpha_{Codata}/(2\pi)$ already accounts for more than 99.8% of the total correction, only higher terms of the development, which reflect interaction processes with much less likelihood, significantly reduce the deviation from the measurement. Remarkably, however, the deviation from the measured value is approximately five times smaller if the Sommerfeld constant α_{geom} derived from a geometrical idea is used in the one-loop term. If α_{geom} is used to calculate the Schwinger term, a value for a_e of ≈0.001159330 results with a deviation from the measured value (≈0.001159652) of only ≈0.028%. What is the cause of the deviation? The geometrized Sommerfeld constant α_{geom} itself or a missing correction δa_e due to additional interactions?

5.4.2 The Shahar Hod limit

Shahar Hod [5.24] has derived a lower bound

Formula 5.6 $\alpha_{Shahar\ Hod} > \ln 3 / 48\pi \approx 1 / 137.26$

on the value of the fine-structure constant. He received this analytical expression by investigating the physics of charged black holes in the framework of a unified quantum theory of gravitation, electrodynamics and thermodynamics. For the geometry of empty space, he used the Reissner–Nordström metric. It is worth noting that the model of an electrically charged black hole gives a numerical value of the Sommerfeld constant $\alpha_{Shahar\ Hod}$, which almost agrees with the geometrized Sommerfeld constant α_{geom} to within ≈150 ppm. Is this a numerical coincidence or is there some relevance to fundamental physics? Similarly to α_{geom}, the numerical fine structure constant $\alpha_{Shahar\ Hod}$ is expressed solely in terms of fundamental mathematical quantities, as required by Richard Feynman and others.

5.5 The universal interaction particle as a consequence of the reference temperature

The local interaction energy between a hole and an electron in vacuum is given by the rest energy of an interaction particle with a thermodynamic mass $m_0 = c^{-2}\, 2\, \pi^{1/3}\, k_B T$ spatially extending over $\Lambda = h\, c^{-1}\, m_0^{-1}$.

Both the interaction particle and the electron are forced to remain spread out in space over a distance given by their Compton length. Both particles would have too much blurring in the impulse if they were located in too small an area. In such a case they would violate the Heisenberg uncertainty principle. Is the almost massless interaction particle the mass quantum postulated in certain theories, which in large quantities move around in space and penetrate all matter practically unhindered?

It is interesting to compare the mass m_0 with the mass m_e of the electron. This comparison corresponds to a comparison of their Compton lengths, since a fundamental mass directly implies a fundamental length and vice versa. If the ratio m_0/m_e is calculated by means of tabulated CODATA values

Relation 5.7 $\lambda_e/\Lambda = m_0/m_e \approx 5.6 \cdot 10^{-7}$

is obtained. The neutral interaction particle with mass m_0 is apparently spread further over space by a factor of $\approx 1.8 \cdot 10^6$, because it is so much lighter than the negatively charged electron.

5.5.1 Localization energy of m_0

A quantum object confined in a one-dimensional, spherical space with radius r is described by a standing wave that must fit into this space. The minimum energy ε that such a micro-object must have does not disappear as in the classical case, but is given by

Formula 5.7 $\varepsilon = h^2\, 8^{-1}\, m^{-1}\, r^{-2}$

as a consequence of the Heisenberg uncertainty principle and the wave-like nature of the micro-object. This lowest energy, which cannot be removed from the sphere, is also called zero-point energy. If $m = m_0$ and $r = \Lambda$ are set, Formula 5.7 yields a characteristic temperature ε/k_B given by

Relation 5.8 $\varepsilon(r = \Lambda;\, m = m_0)/k_B = k_B^{-1}\, h^2\, 8^{-1}\, m_0^{-1}\, \Lambda^{-2} \approx 415.2\ \mathrm{K}\ (\approx 142\ ^\circ\mathrm{C})$

The temperature $\varepsilon(r = \Lambda;\, m = m_0)/k_B$ should actually be observable in experiments. The ratio of the reference energy E_τ to the localization energy $\varepsilon(r = \Lambda;\, m = m_0)$ is independent of natural constants. The ratio is given by

Relation 5.9 $E_\tau / \varepsilon(r = \Lambda;\, m = m_0) = 4\, \pi^{-1/3} = 12\, V_3(c_1) = 2^5\, \alpha_{geom}{}^{1/2} \approx 2.731$

which can be easily deduced by using Formula 5.7 and Relation 2.1. Thus, the localization temperature in units of τ is $\approx 2.731^{-1}$. An experimental reference to this topic can be found at the end of the book in Section 16.5.6.

5.5.2 The electron neutrino

In the beta decay of the neutron, a neutron transforms into a proton, an electron and an antineutrino. The practically unobservable neutrino particle, which has no charge and almost no measurable mass, was introduced to preserve the laws of conservation of energy and momentum in the transformation. In the Standard Model of particle physics, the neutrino is massless and electrically neutral, and it only weakly interacts with matter. The experimental detection of the neutrino is difficult due to the marginal interaction of the particle with matter, and it has only recently been known that the neutrino is likely to have a nonzero mass. If the end of the energy spectrum of the electron is investigated during the beta decay, the energy balance can be used to deduce an absolute (model-independent) neutrino mass m_ν. At present, only an upper limit of 2.2 eV/c^2 can be determined from such experiments for the electron-neutrino mass [5.25]. However, an experiment [5.26] is planned in which the absolute mass of the neutrino can still be measured if the mass is greater than 0.2 eV/c^2. This experiment exploits the tritium (^3H) beta decay at the endpoint of the electron spectrum for the kinematic mass determination of m_ν. Such an absolute mass determination is based solely on kinematic considerations without additional assumptions. Unfortunately, the result of the measurement with a sensitivity of about 0.2 eV/c^2 is not yet available in order to compare the measured mass of the electron neutrino with the mass m_0 of $\approx$0.286 eV/c^2. It becomes interesting, if the mass m_ν is of the order of the mass m_0 of the interaction particle.

5.6 The geometrization of physical quantities

Geometry, that is, the laws of space, is the oldest branch of physics. At a very early stage, rulers were used to measure lengths, area standards to measure areas and liquid volumes to measure volumes. Thus, the geometry has its origin in the measurement. The geometry, originally applied as a natural science, was much later mathematized by axioms. Newton also derived his law of gravity by means of geometric considerations, and his theory therefore reflects a deep connection to geometry.

The physical dimension time can be converted to a length by multiplying it by the constant c. The indication of a distance through a time is actually very common, since on signposts for distances very often times are noted. With the help of the definition of the kilogram resulting from the David Bohm analogy, the kilogram can be replaced by the unit m^3/c^2. Thus, the dimension mass is converted into a volume by multiplying it by the square of the constant c. If this is done consequentially for other units while maintaining the geometric unit of a length, the geometrized units in Table 5.5 for mass, time, energy, charge and temperature follow. Operating only with the dimension of a length, the physical dimensions of mass, energy and temperature become a volume. Since mass, energy and temperature are volumes, all three quantities are geometrically equivalent. They can only be physicalized by means of different factors. Similarly, the physical dimension of charge or force is synonymous with an area. In summary, length means time, area means both force and charge, volume means energy as well as mass or temperature.

In this view, besides the fundamental length, there are exactly three freely selectable, human parameters, namely c, k_B and ε_0, so that it is possible to physically differentiate between the same geometric quantities. The electric field constant ε_0 is a combination of c and μ_0. Both link the electromagnetic units to the mechanical units by means of a force law. With c, k_B and ε_0, all quantities can thus be characterized by geometrical quantities, such as lengths, areas, volumes or dimensionless numbers. Physically important is only that c, k_B and ε_0 are a consistent set of conversion factors between which there are no dependencies. Which values are assigned to the individual conversion factors with uncertainty zero has purely experimental benefits and thus no physical significance. Setting the conversion factors c, k_B and ε_0 all equal to 1 is also correct and, since there is no interdependence, causes no problems. These factors cannot be predicted by any theory, that is, by any mathematical model.

Table 5.5: Geometrized unit system.

Physical quantity	SI unit	Geometrized unit
Length	Meter	m
Time	Second	m/c
Energy	Joule	m^3
Mass	Kilogram	m^3/c^2
Temperature	Kelvin	m^3/k_B
Force	Newton	m^2
Charge	Coulomb	$m^2\,(4\pi\varepsilon_0)^{1/2}$

Note: Geometrized unit system of different physical quantities.

The reference energy E_τ in accordance with Relation 5.2 is proportional to

Relation 5.10 $E_\tau \sim c^{8/9}\, h^{2/3}\, m_e^{1/9}$

The unit of the term $c^{8/9}\, h^{2/3}\, m_e^{1/9}$ is obtained by substituting m/s for the speed of light in vacuum (c), the term kg m^2/s for the Planck constant (h) and kg for the mass of the electron m_e, and is thus given by

Relation 5.11 $[c^{8/9}\, h^{2/3}\, m_e^{1/9}] = m^{20/9}\, s^{-14/9}\, kg^{7/9}$

Since, according to Hypothesis 1.1, the unit kg corresponds to the unit m s^2

Relation 5.12 $[c^{8/9}\, h^{2/3}\, m_e^{1/9}] = m^3$

is obtained as a unit for the reference energy E_T by means of Relation 5.11. This corresponds, as explained at the beginning of this chapter, to a volume and the result of Relation 5.12 is thus consistent with the statement made.

5.7 The universal particle number density

The Mott criterion $n^{1/3}\, a_H^{*} = c_1$ describes the critical point of the Mott metal-insulator transition. This relationship is one of the most important tools for classifying materials into metals or insulators. In doing so, the particle number density n is a decisive parameter.[6] The question arises whether the universal particle number density n of $\approx 3.41 \cdot 10^{18}$ cm^{-3}, found by means of natural constants, is related to the metal-insulator transition or potentially even a quantity that can be measured. Since in the derivation of the number density only natural constants, the number constants 2 and π were used, the number density n has to be material independent and universal.

The classic Mott criterion was given great importance in the past and much experimentation and intensive research were carried out. It turned out that the particle number density cannot be changed experimentally and is neither directly nor easily accessible in crystalline, amorphous or granular materials. Typically, a variety of alloys were examined and the conductivity measured as a function of the stoichiometric composition of a conductive component in an insulating matrix. The influence of the matrix on the measurement cannot be neglected even at very low temperatures. Based on percolation theory the volume fraction is often used to specify the concentration of the conductive component, but this makes it difficult to deduce a particle number density from it.

Another possibility is to selectively change the conductivity in semiconductor systems by doping with different materials and to study in this way the difference between a metal and a nonmetal. The idea of selectively altering the concentration of impurities in an otherwise passive matrix of covalently bonded atoms, that is, in a host lattice of a crystalline solid, is due to N. F. Mott. However, the experimental determination of the critical particle number density n_c, where the transition between a metal and a nonmetal takes place, is also paved with many hurdles. Although many experiments have been performed over the years to determine a universal particle number density in disordered systems of comparable short-range order, no universal behavior was found due to experimental inadequacies in the calibration of impurity

6 The typical atomic number density in solids is $\approx 5 \cdot 10^{22}$ cm^{-3}.

concentrations. However, the materials Si:P and Si:B, which were often used as prototypes for the investigations, gave an experimental impurity density very close to the value of Table 5.1.

H. Stupp and coworkers state a critical particle number density n_c of $\approx 3.52 \cdot 10^{18}$ cm^{-3} for uncompensated silicon Si:P [5.27]. M. J. Hirsch and coworkers [5.28] found for compensated silicon Si:(P,B) in one crystal a critical density of $\approx 3.4 \cdot 10^{18}$ cm^{-3}, in another crystal a critical density of $\approx 4.5 \cdot 10^{18}$ cm^{-3}. Does the particle-number density n of Table 5.1, which depends only on natural constants, be the relevant parameter of the physics of impurities in solids at zero temperature?

5.7.1 The critical current limit

The electrical current density or the electrical flux plays an essential role in electronics, since, for technical and economic reasons, a progressive miniaturization of the conductor paths takes place. At present, the maximum current density in microchips is ≈ 1 MA/cm^2, which in the coming years would be increased to ≈ 3 MA/cm^2. These data are referred to aluminum and aluminum-copper alloys, which are today the most studied materials. Combining fundamental constants and the particle number density n of Table 5.1, a characteristic electric current density or charge current density j_c can be calculated on the grounds of dimensional arguments. This quantity depends only on natural constants and is independent of structural, electrical and diffusive material properties. The critical current limit j_c is given by

Ansatz 5.4 $\quad j_c = c^{1/2}\, h^{3/2}\, \varepsilon_0^{1/2}\, m_e^{-1}\, n^{4/3} \approx 49.6 \text{ MA cm}^{-2}$

Does the current density j_c of ≈ 49.6 MA/cm^2 have an experimental meaning? Is it the material-independent maximum current density at current-impressed measurements before a breakthrough occurs and any three-dimensional material melts, evaporates and finally transmutes into the plasma state? An experimental hint for the existence of j_c can be found in Section 16.5.5.

References

[5.1] F. London and H. London. The electromagnetic equations of the supraconductor. Proceedings of the Royal Society A 149 (1935) 71–88

[5.2] L. De Broglie. Thermodynamique relativiste et mécanique ondulaire. Annales Institut Henri Poincaré 9 (1968) 89–108

[5.3] Robert C. Weast. Handbook of Chemistry and Physics, 55th edition, CRC Press Inc.

[5.4] S. B. Arnason, S. P. Herschfield and A. F. Hebard. Bad metals made with good-metal components. Physical Review Letters 81 (1998) 3936–3939

[5.5] A. F. Hebard and S. B. Arnason. Bad-metal behavior: Exotic physics or a consequence of microstructure? Journal of Superconductivity 12 (1999) 159–162

[5.6] B. Weitzel, A. Schreyer and H. Micklitz. Metal-insulator transition and superconductivity in Bi/Kr composites built from well-defined Bi clusters. Europhysics Letters 12 (1990) 123–128

[5.7] B. Weitzel and H. Micklitz. Superconductivity in granular systems built from well-defined rhombohedral Bi clusters: Evidence for Bi surface superconductivity. Physical Review Letters 66 (1991) 385–388

[5.8] ^{4}He A. Driessen, E. van der Poll and Isaac F. Silvera. Equation of state of solid. Physical Review B 33 (1986) 3269–3288

[5.9] Alexey Bosak and Michael Krisch. Phonon density of states probed by inelastic x-ray scattering. Physical Review B 72 (2005) 224305–224313

[5.10] F. Tuinstra and J. L. Koenig. Raman spectrum of graphite. The Journal of Chemical Physics 53 (1970) 1126–1130

[5.11] R. J. Nemanich and S. A. Solin. First- and second-order Raman scattering from finite-size crystals of graphite. Physical Review B 20 (1979) 392–401

[5.12] Y. Wang, D. C. Alsmeyer and R. L. McCreery. Raman spectroscopy of carbon materials: structural basis of observed spectra. Chemistry of Materials 2 (1990) 557–563

[5.13] Yong P. Chen. http://phys570x.wikispaces.com/file/view/. 570X_lec6_graphene_mechtherm.pdf, queried on August 10, 2013

[5.14] T. R. Hart, R. L. Aggarwal and Benjamin Lax. Temperature dependence of Raman scattering in silicon. Physical Review B 1 (1970) 638–642

[5.15] H. Scheel, S. Khachadorian, M. Cantoro, A. Colli, A. C. Ferrari and C. Thomson. Silicon nanowire optical Raman line shapes at cryogenic and elevated temperatures. Physica Status Solidi B 245 (2008) 2090–209

[5.16] M. R. Winter and D. R. Clarke. Oxide materials with low thermal conductivity. Journal of the American Ceramic Society 90 (2007) 533–540

[5.17] Y. Shen, R. M. Leckie, C. G. Levi and D. R. Clarke. Low thermal conductivity without oxygen vacancies in equimolar. $YO_{1.5} + TaO_{2.5}$ and $YbO_{1.5} + TaO_{2.5}$ stabilized tetragonal zirconia ceramics. Acta Materialia 58 (2010) 4424–4431

[5.18] Z. Qu, T. D. Sparks, W. Pan and D. R. Clarke. Thermal conductivity of the gadolinium calcium silicate appetites: Effect of different point defect types. Acta Materialia 59 (2011) 3841–3850

[5.19] M. J. Feigenbaum. Quantitative universality for a class of nonlinear transformations. Journal of Statistical Physics 19 (1978) 25–52

[5.20] H. E. Nigh, S. Legvold and F. H. Spedding. Magnetization and electrical resistivity of gadolinium single crystals. Physical Review 132 (1961) 1092–1097

[5.21] S. Wurmehl, G. H. Fecher, H. C. Kandpal, V. Ksenofontov, C. Felser, H. Lin and J. Morais. Geometric, electronic, and magnetic structure of Co_2FeSi: Curie temperature and magnetic moment measurements and calculations. Physical Review B 72 (2005) 184434 [9 pages]

[5.22] M. E. Mendoza, F. Donado, R. Silva, M. A. Perez and J. L. Carrillo. Magnetite microcrystals for magneto-rheological fluids. Journal of Physics and Chemistry of Solids 66 (2005) 927–931

[5.23] G. Gabrielse, D. Hanneke, T. Kinoshita, M. Nio and B. Odom. New determination of the fine structure constant from the electron-g-value and QED. Physical Review Letters 97 (2006) 030802 [4 pages]

[5.24] Shahar Hod. Gravitation, thermodynamics, and the fine-structure constant. International Journal of Modern Physics D 19 (2010) 2319–2323

[5.25] Das Mainzer Neutrinomassen Experiment. http://www.physik.uni-mainz.de/exakt/neutrino/de_experiment.html, queried on October 19, 2013

[5.26] Karlsruhe tritium neutrino experiment. http://www.katrin.kit.edu, queried on October 19, 2013

[5.27] H. Stupp, M. Hornung, M. Lakner, O. Madel and H. von Löhneysen. Possible solution of the conductivity exponent puzzle for the metal-insulator transition in heavily doped uncompensated semiconductors. Physical Review Letters 71 (1993) 2634–2637

[5.28] M. J. Hirsch, U. Thomanschefsky and D. F. Holcomb. Critical behavior of the zero-temperature conductivity in compensated silicon Si:(P,B). Physical Review B 37 (1988) 8257–8261

6 The universal energy density

> Given what we know about radioactivity and cosmic radiation, there is no reason anyone can think of why the cosmological constant[1] should not be stupendously large – many orders of magnitude larger than the density of ordinary matter.
> (*Robert B. Laughlin*)[2]

The determination of m_0 and the elementary volume $V = \Lambda\ L\ \lambda_{Lo}$ allows the definition of a homogeneous relativistic mass density rho according to

Definition 6.1 $\text{rho} \equiv m_0 / V \approx 4.43 \cdot 10^{-19}\ \text{kg m}^{-3} \approx 265$ protons per cm^3

The number $4.43 \cdot 10^{-19}$ kg m^{-3} is an enormously small value compared to the mass density of the surrounding matter. Compared to the mean mass density of normal baryonic matter in the observable universe, $4.43 \cdot 10^{-19}$ kg m^{-3} is an enormously large value because it is many orders of magnitude larger than the value of $\approx 5 \cdot 10^{-28}$ kg m^{-3} estimated by the Standard Model of astronomers. Even if the dark, optically unobservable matter – according to today's belief about a factor of six of the baryonic matter – is added, there is still a huge amount of mass per cubic meter missing for the interpretation of the mass density rho. The density of matter in the observable cosmos seems to be a tiny part of the relativistic mass density rho. But, when looking at the total mass density in the galactic disk near the sun, the discrepancy to rho is not very large anymore. After all, the local star density in the vicinity of the sun, derived by means of dynamical models [6.1], roughly amounts to $0.1 \cdot M_{sun}\ \text{pc}^{-3}$, which corresponds to $\approx 7 \cdot 10^{-21}$ kg m^{-3} or about 4 protons per cm^3. Solar particles' investigations in near-solar space by the space probes HELIOS 1 and 2 showed a mean value of the proton density, which better reflects the mass density rho. The space probes detected in situ a proton density of 20–40 cm^{-3} at a distance[3] of ≈ 0.3 AE from the sun [6.2].

If rho is multiplied by the square of the speed of light c^2, the unit kg transforms into the unit J, and the respectful number $(8\pi)^{-1}$ J m^{-3} or $(8\pi)^{-1}$ Pa follows for the energy equivalent rho c^2 (unit J m^{-3}) from the relativistic mass density rho (unit kg m^{-3}). This statement is given by

Relation 6.1 $\text{rho}\ c^2 = (8\pi)^{-1}\ \text{J m}^{-3} \approx 3.98 \cdot 10^{-2}\ \text{J m}^{-3} \approx 39.8$ mPa

1 The cosmological constant is historically a parameter of the form $8\pi\ G\ c^{-2}\rho$ with the unit m^{-2} in Albert Einstein's general relativity equations. One possibility is to interpret the density ρ as the finiteenergy density of the vacuum. However, it is not conclusively clarified today what the cosmological constant means physically, or what value should be chosen for it. The cosmological constant is therefore often synonymous with an energy density of different origin. See also Section 12.1

2 A Different Universe, Basic Books (2005), page 123.

3 An astronomical unit (abbreviated AE) corresponds approximately to the mean distance between earth and sun.

https://doi.org/10.1515/9783110612387-006

Is rho c^2 the relativistic dark energy density of virtual charged particles, that is, of virtual electron-hole pairs that fill in free space, because otherwise the physical vacuum would immediately be filled up by matter? What does carry this uniformly distributed energy in space and why is it there? Dark matter postulated by astronomers, which is invisible because it does not radiate, and manifests itself only in the form of gravitational interactions, cannot be the cause.

6.1 The origin of the electron mass m_e and its definition by the Planck constant h and the speed of light c

Due to the uncertainty principle of Heisenberg, a particle trapped in a volume V has on average a high momentum and a high kinetic energy. This high momentum causes a large pressure similar to the pressure that a gas molecule generates in a container. A particle, that is permanently restricted in its mobility, is constantly under pressure and tries to escape. Formally, such particle movements can also be assigned a temperature.

With the particle number density n (reciprocal volume) and the pressure rho c^2, a characteristic temperature T_{Min} according to

Definition 6.2 $T_{Min} \equiv \text{rho}\, c^2\, n^{-1}\, k_B^{-1} \approx 0.844$ mK

can be calculated in analogy to the equation of state of a classical ideal gas. All particles, regardless of their mass, should contribute the same amount of pressure in analogy to an ideal gas. The temperature T_{Min} is the smallest temperature emanating from combinations of fundamental quantities defined in this book. Definition 6.2 is based solely on dimensional arguments and the value of T_{Min} is universal, since only fundamental constants are needed to calculate it.

Is it the smallest temperature with a meaningful definition that can macroscopically be achieved in thermodynamic equilibrium with a heat bath? According to the third law of thermodynamics or what is equivalent, the heat theorem of W. H. Nernst, there is actually no zero temperature, by whatever process is cooled. Absolute zero is out of reach. In equilibrium, the lowest heat content of the universe cannot be zero according to this theorem. According to Definition 6.2, the heat content or enthalpy could be a cause of the energy density rho c^2 and the particle number density n. The enthalpy would always be conserved because of the law of conservation of energy, that is, because of the first law of thermodynamics.

At sufficiently low temperatures, the thermal properties are no longer dominated by phonons, but by the magnetic moments of the nuclei. The interaction of magnetic moments of atomic nuclei is one of nature's weakest known interaction. The antiferromagnetic order temperature of the solid isotope ^{3}He, consisting of two protons and one neutron, is (0.9 ± 0.02) mK. Is it a coincidence that the antiferromagnetic order temperature, or Néel temperature, of the solid isotope ^3He agrees fairly well with the temperature T_{Min}?

Today, temperatures in the range 0.65–1,350 K are covered by the temperature scale IST-90. From 0.65 K down to 0.9 mK, which corresponds to the Néel temperature of ^3He, the current provisional temperature scale PLTS-2000 applies. Below 0.9 mK there are no official temperature scales defined. Why is T_{Min} the lowest temperature limit of today's temperature measurement?

The thermal De Broglie length λ_{th} is a measure of the length scale on which a matter particle in contact with a heat bath can no longer be localized and gives up its identity in favor of collective behavior. The length λ_{th} is the generalization of the De Broglie length of a single particle ($\lambda = h/p$) for a noninteracting many-body system. For a statistical ensemble of electrons confined in a three-dimensional volume, the thermal length λ_{th} in contact with a heat bath of temperature T_{Min} is given by

Formula 6.1 $\lambda_{th} = h\,(2\pi\, m_e\, k_B T_{Min})^{-1/2} \approx 2.565\ \mu m$

A comparison of the thermal length λ_{th} of the electron with the characteristic length L yields.

Relation 6.2 $\lambda_{th}\, 2^{-1}\, \pi^2 \approx 12.660\ \mu m \approx L \approx 12.677\ \mu m$

using CODATA values. The factor $2^{-1}\,\pi^2$ has been determined heuristically by numerical comparison without a physical background and is, of course, arbitrary at first reading. If, however, the principle is taken as a basis that the factor can only be expressed by possible powers of 2 and π due to symmetry reasons, the choice of the factor no longer appears as arbitrary, because thereby the selection is enormously limited. In addition, if the definition $\alpha = \alpha_{geom} \equiv 2^{-6}\,\pi^{-2/3}$ resulting from Definition 5.3 is used in Relation 6.2 for the Sommerfeld constant α, the numerical agreement is even better

Using only the number constants 2 and π for the factor is supported by the fact that in all of the fundamental quantities derived so far only these two mathematical parameters mysteriously appear as power products. This may indicate that nature actually takes geometry as a guide in defining physical processes. This view is further confirmed by the fact that the factor $2^{-1}\,\pi^2$ corresponds to a geometrical quantity, namely the unit volume of the four-dimensional hypersphere $V_4(1)$. If Relation 6.2 is raised to a hypothesis, that is, if

Hypothesis 6.1 $L\,/\,\lambda_{th} = 2^{-1}\,\pi^2 = V_4(1)$

is required, the constants c, h and m_e are fixed by

Relation 6.3 $c^{5/18}\, m_e^{2/9}\, h^{-1/6} = 2^{5/6}\, \pi^{52/27}$

or

Relation 6.4 $m_{e\_geom} \equiv c^{-5/4}\, h^{3/4}\, 2^{15/4}\, \pi^{26/3}$

which can easily be deduced from Relation 6.3. Can it be that nature is so designed that, assuming proper modeling and accurate measurement, Hypothesis 6.1 is correct for unbound electrons, making m_e predictable without Newton's constant? As the circumference U and the diameter d of a circle are linked through the relationship $U/d = \pi = V_2(1)$, the expression $L/\lambda_{th} = V_4(1)$ also reflects a geometrical situation. The idea that the mass of the electron should be calculable, and that it is not a natural constant had already been formulated by Hendrik Antoon Lorentz. He believed that the electron mass has its origin in the charge of the electron.

The term $c^{5/18}\, m_e^{2/9}\, h^{-1/6}$ has the unit $(kg\ m^{-1}\ s^{-2})^{1/18} = (J\ m^{-3})^{1/18}$. It is dimensionless because of the relationship $kg = m\ s^2$ and has the character of an invariant of the constants h, c and m_e. It follows that the unit of $c^{-5/4}\, h^{3/4}$ must have the unit kg. The geometrized value $m_{e,geom}$ calculated by Relation 6.4 is $\approx 9.066\cdot 10^{-31}$ kg and is approximately 0.5% smaller than the CODATA value obtained from the Rydberg constant R_∞. Half a percent is an astronomical deviation in today's precision physics! How can that be?

The Rydberg constant R_∞ is based on the definition $R_\infty \equiv \alpha^2 m_e c(2h)^{-1}$ and is determined by comparing measured resonance frequencies with theoretical frequencies of hydrogen (H) or deuterium (D), in which a single electron is bound to a nucleus. The theoretical values of the frequencies follow from differences of energy level values, in which the constant occurs as a multiplicative factor. The complicated mathematical formulas for the energy levels of bound electrons are derived from eigenvalues of the relativistic Dirac equation, taking into account nuclear effects and corrections due to quantum electrodynamics (QED).[4] In short, the tabulated electron mass $m_{e,Codata}$ is not directly measured, but is an indirect quantity derived from models of energy levels and strongly correlated [6.3] with the fine-structure constant α_{Codata}. If the mass m_e is fixed by Relation 6.4, the electron mass m_e cannot be a free parameter of the Rydberg constant R_∞ anymore. Resonance frequencies have to be interpreted differently not only because of $m_{e,geom}$ but mainly because of α_{geom}. This topic will be revisited in Chapter 15.

By the axiomatic definition that the dimensionless ratio L/λ_{th} should correspond to the number constant $2^{-1}\,\pi^2$ or, from a geometrical point of view, the unit volume of

4 In QED, the electron mass is a free parameter, which is predicted by the equations to be infinite and only gets a finite value by renormalization. All divergences of QED are absorbed by the renormalization in the electron mass! The same facts apply to the charge.

the four-dimensional hypersphere, the electron mass m_e, which together with other natural constants determines the atomic scale of $\approx 10^{-10}$ m, becomes a property of matter that can be reduced to the constants h and c. In this view, Planck's constant h and the speed of light in vacuum c are more fundamental than the mass of the electron, which thereby loses its independence. If the number constant $2^{-1}\,\pi^2$ is a determinant of matter, it is probably based on a statistical principle that remains puzzling and certainly requires a deeper explanation. There is no clear objective evidence to support the previous view, and it is certainly of major concern to axiomatically postulate that the mass of the electron is based on a collective effect of many electrons and is created dynamically by thermal movements.[5]

Is the electron mass a basic physical necessity, because an ensemble of negatively charged electrons would otherwise be unstable because of the Coulomb repulsion? Or could Hypothesis 6.1 be concerned with the classically unpredictable ambiguity of the electron spin? The spin was first introduced on purely phenomenological ground as the fourth degree of freedom of the electron to explain the Zeeman effect in hydrogen. To associate Hypothesis 6.1 with magnetism is not so far-fetched, since the length L will also play an important role in Section 11.6 as a radius of magnetic interaction.

Probably the most important argument for Hypothesis 6.1 is that due to the geometrical idea expressed by Hypothesis 6.1, the anomalous magnetic moment of the electron matches the measured value up to the 13th decimal place after the comma. This excellent agreement would not be possible without Relation 6.4 following from Hypothesis 6.1. Detailed information on these relationships can be found in Section 11.7.

The thermal De Broglie wavelength given by Formula 6.1 is a quantity used in statistical physics, suggesting that the macroscopically measurable mass of the electron has something to do with statistics. Does this lead to the conclusion that the electron mass is purely quantum mechanical in nature and has little to do with ordinary gravitational mass? The thermal De Broglie length reflects the quantum nature of matter, that is, the concept of wave particle duality and connects classical particle physics and quantum mechanical wave physics, as it associates a particle with a wavelength. The connection of the mass with quantum mechanics is not clarified today or experimentally verified. A theory for the internal structure of the electron is also unknown, and attempts to explain the electron mass with electromagnetic interactions have all led to contradictions.

The definition of the electron mass by means of Relation 6.4 determines the atomic mass unit u and thereby also the Avogadro constant $N_{Avogadro}$, since m_e and u are related by $m_e = A_r(e)$ u. The dimensionless relative atomic mass $A_r(e)$ can be measured very accurately in a Penning trap by comparing cyclotron frequencies

5 Does this possibly imply that the Schrödinger equation does not describe a single electron but a statistical ensemble of electrons, that is, a system of many particles?

Definition 6.3 $L_{large} \equiv c\,(G\,rho)^{-1/2} \approx 5.52 \cdot 10^{22}$ m $\approx$ 1.79 Mpc

can be calculated on the grounds of dimensional arguments. It is the largest length that results from a combination of natural constants with rho and implies a finite volume of space. This actually makes sense, since physics needs finite and measurable quantities so that a relation to reality can be established. Does this mean that the interaction between masses becomes weaker or even unstable at astronomical distances? Could the gravitational effect in near cosmic space be limited to the length scale L_{large}? It is a fact that the structure or the cohesion of a galaxy cannot be explained with classical Newtonian laws alone.

All systems of stars or galaxies are organized in clusters or groups. Our solar system is part of the Milky Way, which together with the Andromeda galaxy belongs to the local group of the nearby cosmos. The two spiral galaxies Milky Way and Andromeda are the most massive members of the local group. All those galaxies belong to a group, of which the astronomers assume that they are gravitationally bound and not a random geometric structure. The local group is considered a typical representative of other galaxy groups. As a simple criterion of the membership to the local group, the distance of a galaxy to the center of mass is often used. If the distance is less than ≈1.5 Mpc, the probability is high that the galaxy belongs to the local group [6.8]. Is it a coincidence that the criterion of membership is on the order of L_{large}?

The length scale of the observable universe is ≈4,200 Mpc. This length of cosmology, also called the Hubble length R_{Hubble}, is obtained by dividing the speed of light by the Hubble [6.9] parameter H_0 of (72 ± 8) km s^{-1} Mpc^{-1}. Why is the length $R_{Hubble} \equiv c/H_0$ of ≈4,200 Mpc orders of magnitude greater than the intergalactic length scale L_{large} of ≈1.79 Mpc? How are these two parameters related? Could it be that gravity on cosmological scales, that is, at distances larger than L_{large}, no longer operates? Actually, such an assumption is not a contradiction, since the validity of Newton's laws on the scale of galaxies has never been experimentally verified.

If a test particle of mass m is approached to a mass M of size R

Condition 6.1 $G\,M\,m\,R^{-1} = m\,c^2$

must hold, if no energy is required for the approach and the formation of the test particle. If the mass M is given by $M = (4/3)\,\pi\,R^3\,rho$, Condition 6.1 allows the calculation of the sphere radius R, which defines the three-dimensional volume $(4/3)\,\pi\,R^3$, within which the confinement of any particle exactly costs its rest energy. The sphere radius R is given by

Relation 6.8 $R_{horizon} = (3/4\pi)^{1/2}\,c\,(G\,rho)^{-1/2} = (3/4\pi)^{1/2}\,L_{large}$

from which

Relation 6.9 $R_{horizon} \approx 0.489\ L_{large} \approx 2.69 \cdot 10^{22}\ m \approx 873\ kpc$

can be derived. Through this simple energy consideration of quantum physics, the intergalactic length L_{large} takes on a different physical meaning. The largest length is roughly the diameter of a finite spherical cavity of radius $R_{horizon}$ from which each particle needs more than its rest energy to escape. Thus, in a statistical sense, the effect of gravitation is limited to a maximum distance and all processes are only possible within a causally connected particle volume of a homogeneous mass density determined by the particle horizon.

6.4 An estimate of the upper limit of the neutrino mass in the beta decay

The radiation law of Max Planck suggests a zero-point energy of (hf)/2 of the electromagnetic field in vacuum, since a temperature-independent parameter arises in the derivation of the formula. This residual energy, called quantum fluctuations by quantum theory, exists virtually and is part of the vacuum that is not really empty when all visible matter such as solids, liquids and gases are removed. The question of the energy density of the vacuum is currently considered unanswered. On the one hand, theoretical values on the order of $10^{+115}\ J\ m^{-3}$ are derived from the vacuum fluctuations by quantum electrodynamics and quantum field theory. On the other hand, measurements in astrophysics show a vacuum density of the order of about $10^{-9}\ J\ m^{-3}$ due to the assumed expansion of the observable universe. There is a huge discrepancy between these two values and neither is close to rho c^2 of $\approx 3.98 \cdot 10^{-2}\ J\ m^{-3}$.

For electromagnetic waves, the mode or particle number density (unit m^{-3}) within the frequency interval df is given by

Formula 6.2 $\delta n = 8\pi f^2 c^{-3}\ df$

The energy density results from summing up all frequencies according to

Relation 6.10 $\varepsilon / V = \int \delta n\ (h\ f/2) = 4\pi\ h\ c^{-3} \int f^3 df$

According to quantum electrodynamics, the frequencies in Relation 6.10 can range from zero to infinity when summing up all vibrational zero-point energies. Short wavelength vibrations have high frequencies and thus contribute more to the energy density. But summing up all frequencies results in an infinite integral and thus an infinite amount of energy in the field, which does not make sense physically. If, in order to solve the convergence problem in the calculation of the energy density ε/V, the divergent improper integral over all zero-point frequencies is replaced by an integral with integral limits from zero to the cutoff frequency $f_c = c/L_c$,

Relation 6.11 $\varepsilon / V = h\, c\, \pi\, L_c^{-4}$

results. In this way, high-energy contributions are omitted, but the idea of localizing the energy in a field is maintained. Vibrations with wavelengths smaller than L_c should not be used because the modes can no longer be considered as independent.

A small elementary length, which can be formed solely from natural constants, is the Planck length L_{Planck} explained by

Definition 6.4 $L_{Planck} \equiv h_{bar}^{1/2}\, G^{1/2}\, c^{-3/2}$

It has the value of $\approx 1.6 \cdot 10^{-35}$ m. However, their physical meaning is unclear and controversial. If it is assumed that each wavelength has a lower bound at the Planck length L_{Planck} or each frequency a corresponding upper bound, a value of $\approx 9.1 \cdot 10^{+114}$ J/m^3 is obtained from Relation 6.11 for the energy density ε/V. This limit is considered very skeptical by most physicists, as it appears to be much too high. It is probably just an upper limit on the vacuum energy density of space filled with virtual particles.

If ε/V is equated to the energy density rho $c^2 = (8\pi)^{-1}\ J/m^3$ according to

Ansatz 6.2 $\varepsilon / V = h\, c\, \pi\, L_c^{-4} = \text{rho}\ c^2$

an equation for a cutoff length L_c can thereby be obtained. From

Equation 6.1 $(8\pi)^{-1} = h\, c\, \pi\, L_{c,rho}^{-4}$

a value of ≈ 1.99 μm for $L_{c,rho}$ defined by

Definition 6.5 $L_{c,rho} \equiv 2^{3/4}\, \pi^{1/2}\, (hc)^{1/4}$

can be calculated. The corresponding cutoff frequency $f_{c,rho} \equiv c/L_{c,rho}$ amounts to $\approx 1.5 \cdot 10^{14}$ Hz, which corresponds to a wave number of $\approx 5{,}025\ cm^{-1}$.

Molecular hydrogen (H_2) has the highest vibrational energy of all molecules and has the fundamental vibrational transition at $\approx 4{,}161\ cm^{-1}$. There is apparently no molecular fundamental frequency greater than the cutoff frequency $f_{c,rho}$. Does this suggest that the frequency $f_{c,rho}$ (or the corresponding length $L_{c,rho}$) could in fact have a universal meaning? In any case, Ansatz 6.2 provides a physically more meaningful cutoff frequency than the Planck frequency. From $f_{c,rho}$ a timescale of ≈ 6.6 fs results.

Hydrogen in its molecular form has other notable properties. Its binding energy is (104.207 ± 0.001) kcal/mol [6.10] or ≈4.520 eV, which corresponds closely to one-third (4.52/13.6 ≈ 0.332) of the ionization energy of atomic hydrogen. Fractional numbers also seem to play a role in the bond distance which is ≈0.7414 Å or about 7/5 (0.7414/0.5292 ≈ 1.401) in units of the Bohr radius a_H.

A numerical comparison of the cutoff length $L_{c,rho}$, calculated by Equation 6.1 using CODATA values, with the Compton length Λ of the mass quantum m_o is visualized by

Relation 6.12 $\Lambda / L_{c,rho} \approx 2.177$

The explicit dependence of the ratio $\Lambda / L_{c,rho}$ on natural constants is given by

Relation 6.13 $\Lambda / L_{c,rho} = c^{-5/36}\, h^{1/12}\, m_e^{-1/9}\, \alpha^{-1/3}\, 2^{11/12}\, \pi^{-5/54}$

Setting $\alpha = \alpha_{geom}$ and $m_e = m_{e\_geom}$

Relation 6.14 $\Lambda / L_{c,rho} = 2^{5/2}\, \pi^{-5/6} \approx 2.179$

results. By transforming the ratio $\Lambda / L_{c,rho}$, a relationship between the cutoff energy $hf_{c,rho}$ and the energy $m_o c^2$ can be obtained. This connection is represented by

Relation 6.15 $(h\, f_{c,rho})/(m_o c^2) = \Lambda/L_{c,rho}$

Due to

Relation 6.16 $h\, f_{c,rho} = (\Lambda/L_{c,rho})\, m_o c^2$

following from Relation 6.15, the cutoff energy is limited by the rest energy of a particle of mass $(\Lambda/L_{c,rho})m_o$. Does the mass $(\Lambda/L_{c,rho})m_o$ correspond to the neutrino mass m_ν? Let the mass m_ν be defined by

Definition 6.6 $m_\nu / m_o = \Lambda / L_{c,rho}$

From Equation 6.1 and Relation 6.15, by simply transforming,

Relation 6.17 $\pi\, h^{-3}\, c^5\, m_\nu^4 = (8\pi)^{-1} = rho\ c^2$

follows. It shows, together with Equation 6.1, that $L_{c,rho}$ is the Compton length of m_ν. Numerically, Definition 6.6 yields a value of ≈0.62 eV/c^2 for the upper limit of the neutrino mass m_ν.

6.5 The gravitational self-energy and the connection of the Newton constant G to h and c

The total energy of a system of charges q is the sum of the Coulomb terms resulting from the mutual interaction of each charge pair. For a homogeneous charge distribution, the sum of all possible interaction terms can be replaced by an integral. The total energy or self-energy of a sphere of radius r uniformly filled with the total charge Q is well defined [6.11] and is given by

Formula 6.3 $\varepsilon_{Coulomb} = (3/5)\, Q^2\, (4\pi\varepsilon_o\, r)^{-1}$

Because of the correspondence relationship $Q^2/(4\pi\varepsilon_0) \rightarrow G\, M^2$, the total energy of a sphere of radius r homogeneously filled with the total mass M is given by

Formula 6.4 $\varepsilon_{grav} = G\, \Sigma m_i m_j / r_{ij} = (3/5)\, M^2\, G\, r^{-1}$

The radius r of this gravitationally bound sphere filled with gravitating particles shall be defined according to

Formula 6.5 $r = h_{bar}/(Mc)$

specifying the length scale of a local process in the microcosm with mass M and r^{-3} particles per length cube. With this particle concept, a characteristic, gravitational energy density can be specified. The concept is based on an idea of Zeldovich [6.12] who interpreted the gravitational energy of the vacuum as a dynamic interaction of virtual particles separated by a mean distance of $h_{bar}/(Mc)$. Ya. B. Zeldovich chose the mass of the proton for the mass M without specifying the reason.

Combining Formulas 6.4 and 6.5,

Relation 6.18 $\varepsilon_{grav}/V = \varepsilon_{grav}/r^3 = (3/5)\, (2\pi)^4\, M^6\, G\, c^4\, h^{-4}$

is obtained for the energy density ε_{grav}/V, which could be responsible for how gravity works at microscopic scales. If the gravitational energy density ε_{grav}/r^3 is equated to the energy density rho c^2 or $(8\pi)^{-1}$ J m^{-3} in analogy to Ansatz 6.2,

Equation 6.2 $(8\pi)^{-1} = (3/5)\, (2\pi)^4\, m_{grav}{}^6\, G\, c^4\, h^{-4}$

for the gravitational mass $M = m_{grav}$ is obtained. From Equation 6.2, the local mass scale m_{grav} can be calculated with CODATA values for G, c and h. Since energy is equivalent to mass due to Albert Einstein's equivalence principle, m_{grav} can be considered as the gravitational emergence of the energy density rho c^2. If the value determined from Equation 6.2 is normalized with the CODATA value of the electron mass m_e,

still play a central role? Is our world only a part of a set of higher, additional dimensions that control physics much more than we think?

The reduced Compton length of the charged pion of mass $m_{pion\_charged}$ is defined by

Definition 6.8 $r_{pion\_charged} \equiv h_{bar}/(m_{pion\_charged}\ c)$

to which reference is made in later chapters, for example, in Relation 9.12. The numerical value of $r_{pion\_charged}$ is ≈1.414 fm.

The mean charge radius $<r^2_\pi>^{1/2}$ of the pion averaged from various experiments [6.14] is (0.672 ± 0.008) fm. It is noteworthy that this experimental average agrees very well with the value of $3r_{grav}$ of ≈0.670 fm, which can be calculated using the constants G, c and h. Is it a numerical coincidence that the mean charge radius $<r^2_\pi>^{1/2}$ in units of r_{grav} amounts to ≈3? Is there a reason for the number three?

6.5.3 A tour to the world of condensed matter

The following annotations have nothing to do with gravitational interactions at all. The topic is discussed here because the Bohr radius a_H plays a crucial role similar to Hypothesis 6.2.

By pressure, distance-dependent interactions can be influenced in a controlled manner, and by observing macroscopic properties microscopic mechanisms can be explored. Phase transitions, induced by pressure, are due either to magnetic, electronic or structural transformations. Hypothesis 6.2 establishes a relationship between the microscopic temperature τ, the Bohr radius a_H and the mass m_{grav} of a particle. Could it be that τ and a_H are related in a different way than by Hypothesis 6.2? Using the energy E_τ and the volume a_H^3, it is easy to deduce a pressure given by

Relation 6.26 $P = E_\tau/a_H^3 \approx 105.6$ GPa

on the grounds of dimensional arguments like Definition 6.2.

On the classical Mott insulator manganese oxide (MnO), electrical conductivity measurements [6.15] at room temperature under high, variable pressures revealed a steady decrease in resistance up to 90 GPa, followed by a dramatic decrease by a factor of 10^5 between 90 and 106 GPa. Temperature cycles showed insulating behavior (negative TCR) at 87 GPa and metallic behavior (positive TCR) at 106 GPa. Using x-ray emission spectroscopy, C. S. Yoo and coworkers [6.16] localized the same transition at (105 ± 5) GPa. At this pressure, they found a collapse of volume and a significant loss of magnetic moments. It is believed that at (105 ± 5) GPa the electrons abruptly change from an extended to a localized state. Is there a relationship between the measured pressure of (105 ± 5) GPa and the pressure of ≈105.6 GPa calculated with Relation 6.26?

References

[6.1] O. Bienaymé, A. C. Robin and M. Crézé. The mass density in our galaxy. Astronomy and Astrophysics 180 (1987) 94–110
[6.2] Michael Stix. The sun: An introduction. Springer-Verlag Berlin, 1989
[6.3] Peter J. Mohr and Barry N. Taylor. Adjusting the values of the fundamental constants. Physics Today 54 (2001) 29–34
[6.4] Dean L. Farnham, Robert S. Van Dyck and Paul B. Schwinberg. Determination of the electron's mass and the proton/electron mass ratio via Penning trap mass spectroscopy. Physical Review Letters 75 (1995) 3598–3601
[6.5] F. Zwicky. The molecular hydrogen content of the universe. Publications of the Astronomical Society of the Pacific 71 (1959) 468–469
[6.6] NIST Standard Reference Database Number 69. http://webbook.nist.gov/chemistry, queried on October 7, 2011
[6.7] H. M. Goldenberg, D. Kleppner and N. F. Ramsey. Atomic hydrogen maser. Physical Review Letters 5 (1960) 361–362
[6.8] Ch. J. Pritchet and S. Van Den Bergh. The luminosity distribution of local group galaxies. The Astronomical Journal 118 (1999) 883–888
[6.9] W. L. Freedman, B. F. Madore, B. K. Gibson et al. Final results from the Hubble space telescope key project to measure the Hubble constant. The Astrophysical Journal 553 (2001) 47–72
[6.10] Robert C. Weast. Handbook of Chemistry and Physics, 55th edition, CRC Press Inc.
[6.11] R. P. Feynman, R. B. Leighton, and M. Sands. Elektromagnetismus und Struktur der Materie. Feynman Vorlesungen über Physik, Band 2.
[6.12] Ya. B. Zeldovich. Cosmological constant and elementary particles. Pis'ma Zh. Eksp. Teor. Fiz. 6, 883–884. Journal of Experimental and Theoretical Physics 6 (1967) 316–317
[6.13] J. F. Crawford, M. Daum, R. Frosch et al. Precision measurements of the pion mass difference. Physical Review D 43 (1991) 46–58
[6.14] Particle data group. http://pdg.lbl.gov, queried on October 8, 2011
[6.15] MnO. J. R. Patterson, C. M. Aracne, D. D. Jackson, V. Malba and S. T. Weir. Pressure-induced metallization of the Mott insulator. Physical Review B 69 (2004) 220101 [4 pages]
[6.16] MnO. C. S. Yoo, B. Maddox, J. H. P. Klepeis, V. Iota, W. Evans and A. McMahan. First-order isostructural Mott transition in highly compressed MnO. Physical Review Letters 94 (2005) 115502 [4 pages]

7 Universal parameters of collective vibrations of the plasma

In addition to the three states of aggregation (solid, liquid and gaseous), matter can also form a plasma state, which can cause collective phenomena via forces acting between the charge carriers. This state is described by the plasma frequency f_p and the derived quantities plasma length L_p and plasma energy E_p. The plasma frequency is an alternative measure of the mean density of the free charge carriers per unit volume and is defined by

Definition 7.1 $\omega_p \equiv 2\pi \cdot f_p = q_e\, \varepsilon_0^{-1/2}\, (n^*/m^*)^{1/2}$

From this

Definition 7.2 $L_p \equiv c/f_p$

Definition 7.3 $E_p \equiv h_{bar}\, \omega_p = h\, f_p$

derive. Setting $m^* = m_e$ and $n^* = n$ ($\approx 3.41 \cdot 10^{18}$ cm^{-3}) from Table 5.1, the values listed in Table 7.1 for f_p, L_p and E_p, which depend only on natural constants, are obtained.

Table 7.1: Universal parameters of the collective plasma.

Parameter	Value	Unit	
f_p	16.6	THz	
L_p	18.1	µm	
E_p	68.6	meV	≈796 K ≈ 523 °C

Note: The relevant parameters of the collective plasma: All values can solely be calculated by means of natural constants.

7.1 Amorphous germanium

R. C. Chittick [7.1] investigated the electrical and optical properties of amorphous silicon and germanium and compared the results with crystalline materials. The infrared absorption spectrum of amorphous germanium has two strong peaks at ≈0.23 and ≈0.07 eV, which disappear when the amorphous material is heated to 780 K for about 1 h, because the amorphous material crystallizes. These bands do not change in position or in strength when the films are cooled to 77 K. The two peaks therefore seem to be associated with the amorphous state. The spectrum of silicon is completely flat. Is the absorption peak at ≈0.07 eV determined by the plasma energy E_p of ≈68.6 meV?

https://doi.org/10.1515/9783110612387-007

7.2 Si:P near the metal-insulator transition

A. Gaymann and coworkers [7.2] investigated the far-infrared reflection and its temperature dependence on uncompensated Si:P near the metal-insulator transition. The measurements covered a spectral range between 2.5 meV (20 cm^{-1}) and 620 meV (5,000 cm^{-1}). All optical reflectance spectra $R(h_{bar}\omega)$ exhibit as a function of $h_{bar}\omega$ a well-defined reflectance minimum, located at ≈70 meV for doping concentrations in the vicinity of the metal-insulator transition ($n_c \approx 3.5 \cdot 10^{18}$ cm^{-3}). With increasing doping with respect to the metal-insulator transition (metallic samples), the minimum shifts to higher energies. With decreasing doping (insulating samples), a shift to lower energies occurs. The position of the reflectance minimum is temperature independent at the metal-insulator transition, because it remains at ≈70 meV for all temperatures between 10 and 300 K. Is the reflectance minimum at about 70 meV determined by the plasma energy E_p of ≈68.6 meV as in the previous example?

References

[7.1] R. C. Chittick. Properties of glow-discharge deposited amorphous germanium and silicon. Journal of Non Crystalline Solids 3 (1970) 255–270

[7.2] A. Gaymann and H. P. Geserich. Temperature dependence of the far-infrared reflectance spectra of Si:P near the metal-insulator transition. Physical Review B 52 (1995) 16486–16493

8 The duality relation or the connection between microcosm and macrocosm

One does not get anywhere simply by going over the successes again and again, whereas by talking over the difficulties people can hope to make some progress.
(*Paul Dirac*)[1]

The mean free path is the mean of the distance a particle can travel without interacting with other particles. Interactions require that particles have a finite size and are not considered as idealized, mathematical points. The particle image has the mathematical advantage that no singularities are generated.

On the grounds of dimensional arguments, and, if systems are considered that are in equilibrium with a heat or particle bath of number density n, then for two arbitrary lengths L_1 and L_2

Relation 8.1 $L_2(L_1) = (n \cdot L_1^2)^{-1}$

must apply. Physically, the length square L_1^2 can be interpreted as an impact cross section. Mathematically, the area L_1^2 is transformed into the length $L_2(L_1)$ by means of the particle number density n, which corresponds to a reciprocal volume. This projection of two dimensions onto one dimension reflects a geometrical issue associated with a constant, reciprocal volume. In the physical sense, characteristics of the macroscopic or microscopic physical world with characteristic length scales are transformed into a physical world with corresponding length scales, that is, a long distance action is feeding back to a short distance action. The particle number density n is the link between these two worlds. In a sense, this corresponds to a reductionistic view that associates large things to more fundamental, smaller things, and implies that macroscopic phenomena ultimately depend on a finite number of fundamental building blocks and their interactions. Or simply formulated, the physics of the large follows from the physics of the small and vice versa. The exploitation of the large and the small by means of Relation 8.1 amounts to an idea of David Bohm, who took the view that every point in space is like the whole cosmos. In contrast to conventional theories, Relation 8.1 generates no singularities, that is, lengths or energies are bounded upward and downward. Whether Relation 8.1 is a useful working tool for why there are small and grand things in the universe is the subject of subsequent sections. In any case, it opens an alternative way to study the as yet unexplained relationship between the micro- and the macrocosm.

1 The evolution of the physicist's picture of nature, Scientific American 208 (1963) 45–53.

https://doi.org/10.1515/9783110612387-008

In Section 5.1 it was shown that the length λ_{MI} can be generated by a transformation from the length $L_1 = \lambda_{Lo}$. Similarly, for the lengths $L_1 = \Lambda$, L, L_p and a_B together with the particle number density n of $\approx 3.41 \cdot 10^{18}$ cm^{-3} corresponding lengths $L_2(L_1)$ can be calculated according to Relation 8.1. Since both n and all lengths L_1 mentioned earlier depend only on natural constants, Relation 8.1 yields lengths L_2, which depend only on natural constants as well. In short, Relation 8.1 requires no new parameters. In Table 8.1, the length hierarchy is listed, which results from different lengths L_1, ordered by size. $L_{equal} = L_2(L_{equal})$ yields a value for L_{equal} of $\approx$6.64 nm.

Table 8.1: Interaction lengths.

L_1		$L_2(L_1)$	
L_p	18.1 μm	0.90 fm	$= r_e/\pi$
L	12.7 μm	1.82 fm	
Λ	4.3 μm	15.6 fm	
λ_{Lo}	21 nm	0.66 nm	$= \varepsilon_0^{-1} c^{-2} q_e^2 \cdot m_e^{-1} \alpha^{-2}$
			$= 2 h c^{-1} \alpha^{-1} m_e^{-1}$
			$= \lambda_{MI}$
L_{equal}	6.64 nm	6.64 nm	$= L_{equal}$
a_B	2.52 nm	46.3 nm	

Note: For various lengths L_1, corresponding lengths $L_2(L_1)$ calculated by Relation 8.1 are shown. For certain lengths L_2, existing identities are listed in a separate column as well.

The length measures $L_2(L_1)$ listed in Table 8.1 cover a range from the nanoscale down to the femtoscale. Are these lengths characteristic parameters for describing the atomistic dynamics and the physics of the nuclei? Which boundary assigns the invariant length L_{equal} of $\approx$6.64 nm? Is it the dividing line between micro and macro?

The quantities L_p^2, L^2 and Λ^2 are areas or cross sections and can be considered as a measure of the strength of an interaction. Strong interactions give large areas and weak interactions produce small areas. If the strongest interaction is normalized to one, the values in Table 8.2 are obtained for the relative strengths of the three largest cross sections.

Table 8.2: Cross sections.

L_1^2	Cross section (cm²)	Rel. strength	Range L_2 (fm)	Mass (MeV c⁻²)
L_p^2	$3.27 \cdot 10^{-6}$	1	0.90	220.0
L^2	$1.61 \cdot 10^{-6}$	0.49	1.82	108.3
Λ^2	$1.88 \cdot 10^{-7}$	0.057	15.6	12.64

Note: Relative strengths of the three largest cross sections. The energy corresponding to the range $L_2 = (n \cdot L_1^2)^{-1}$ is represented by $h_{bar}\, c\, L_2^{-1}$. The corresponding mass is obtained by dividing this value by the speed of light squared c^2.

The heaviest known element, which dates back to the time of the origin of the solar system and has not yet completely disintegrated, is the long-lived plutonium isotope ^{244}Pu with 94 protons and 150 neutrons. Its half-life is ≈$8 \cdot 10^7$ years. This transuranium isotope, which occurs in minute amounts, was first found in the Californian ore bastnasite. The nuclide ^{244}Pu was shown to be primordial[2] in 1971 by D. C. Hoffman and coworkers using mass spectrometry [8.1].

According to current doctrine, the nuclear diameter of ^{244}Pu is about $2 \cdot (1.25 \pm 0.05) \cdot 244^{1/3}$, or (15.6 ± 0.6) fm due to the experimental rule $R_{nucleus} \approx r_0 \cdot A^{1/3}$. Although the nuclear diameter is a quantity that cannot be measured very accurately because of the blurring of the edge, it is amazing that the nuclear diameter of the heaviest, naturally occurring element ^{244}Pu is approximately equal to the length $L_2(\Lambda)$.

The nickel isotope ^{62}Ni with 28 protons and 34 neutrons shows the largest binding energy per nucleon of ≈8.8 MeV and is thus the most stable nuclide. Its nuclear diameter is about $2 \cdot (1.25 \pm 0.05) \cdot 62^{1/3}$ fm or ≈ (9.9 ± 0.4) fm according to the current doctrine. Interestingly, this corresponds to $\approx L_2(\Lambda) \cdot (2/\pi)$ or ≈9.9 fm. The factor $2/\pi$ and the characteristic length $L_2(\Lambda)$ will again play a role in Section 8.2, and the coincidence, that the nuclear diameter of the nuclide with the largest mass defect per nucleon is approximately $L_2(\Lambda) \cdot (2/\pi)$, must be seen in this context.

8.1 The mean charge radius of the proton and the nucleon density of extended nuclei

The proton is the nucleus of the hydrogen isotope ^1H and the most abundant nucleus in the observable universe. Many experimental findings suggest that the nucleus has almost a defined surface and therefore the definition of a mean nuclear radius makes sense. The measurement of the charge radius, however, is sensitive to the measurement method, and it is therefore not surprising that there is a large spread in the literature values. Even when the same measuring principle is used, contradictory values in the extensive literature can be found.

Until a few years ago the charge radius of the proton was determined in two ways. Either the nucleus was bombarded with electrons and the radius calculated by means of the deflection of the electrons, or the radius was determined indirectly by a model, that is, computed from the hyperfine splitting of the hydrogen atom by means of complicated formulas of quantum electrodynamics (QED). In the last method, the charge radius is in principle nothing more than a free parameter in an abstract formalism. It seems clear that protons are no point charges, but are somehow smeared out, have a structure and can strongly interact with pions.

2 The half-life has to be greater than ≈50 million years due to the age of the earth of ≈4.6 billion years, to be sure that a nuclide originates from the time of the origin of the solar system.

Ingo Sick [8.2] studied a great amount of world data on elastic electron-proton scattering experiments to determine the proton charge rms-radius $<r^2_{proton}>^{1/2}$. By analyzing the data he found a mean charge radius of (0.895 ± 0.018) fm. It is interesting that the characteristic length $L_2(L_p)$ coincides very well with the proton charge rms-radius obtained by Ingo Sick. Using CODATA values for all constants

Relation 8.2 $<r^2_{proton}>^{1/2} \approx L_2(L_p) = (n\, L_p^2)^{-1} = r_e\, \pi^{-1} \approx 0.897$ fm

can be derived. The quantity r_e denotes the classical electron radius, which is defined without any physical meaning as a pure arithmetic parameter of length given by the expression $q_e^2/(4\pi\varepsilon_0 c^2 m_e)$. If r_e is multiplied by the geometric factor π^{-1}, it obviously amounts to a value of the proton charge rms-radius, which agrees fairly well with the value that Ingo Sick determined from data of the particle physicists. The value compiled by CODATA, which is obtained mainly from precision spectroscopic data of hydrogen by complex QED calculations, is significantly smaller and amounts to approximately 0.877 fm.

Because of the significant disagreement, P. G. Blunden and Ingo Sick [8.3] investigated the influence of two-photon exchange processes on the proton charge rms-radius extracted from data on electron-proton scattering experiments. They found that the changes are small and do not help explain the discrepancy between the experimental value and that calculated from the hyperfine structure of hydrogen. On the contrary, the discrepancy became even larger, since after the two-photon correction the experimental rms-radius of the particle physicists was now (0.897±0.018) fm. Interestingly, however, the deviation from $L_2(L_p)$ becomes smaller.

Recent calculations of the proton charge rms-radius from spectroscopic data on muonic hydrogen make the confusion even larger, as Randolf Pohl and coauthors [8.4] calculated ≈0.842 fm. Although precision spectroscopic methods are very accurate in terms of measurement, the calculations of bound-state QED needed to determine the proton charge rms-radius are much more complicated than the traditional calculation by means of cross sections of scattering processes.

What is the cause of the discrepancy? Are there errors in the calculations or must even QED be questioned, which is considered to be very well proven? The length $L_2(L_p)$ clearly favors the scattering data of the nuclear physicists and suggests that in fact something could be wrong with the QED-based spectroscopic calculations of the atomic physicists. An alternative explanation of the proton charge rms-radius is presented in Section 14.7.

8.1.1 The nucleon density of extended nuclei

It is known from scattering experiments on many elements with electrons and neutral pions that the number density of nucleons inside heavy nuclei is independent of the mass number and amounts to about 0.17 nucleons per fm^3. Each nucleon inside the nucleus apparently occupies on average a cube of ≈6 fm^3. The mean distance of the nucleon interaction is thus calculated as $(1/0.17)^{1/3}$ fm. The resulting value of ≈1.8 fm is interestingly equivalent to ≈$L_2(L)$. Is the length $L_2(L)$ the relevant length scale of the nucleon density?

8.2 The parameterizing of nuclear binding energies and the associated energy scale

The binding energy of an interacting nucleon system of protons and neutrons is expressed in the mass deficit δm, which states the difference of the free masses of the number of protons and neutrons to the actual total mass of the nucleon. The result of a linear regression [8.5] of nuclear binding energies B based on an experimental data set of 2,932 isotopes as a function of the number of protons n_p and the number of neutrons n_n gave

Relation 8.3 $B(n_n, n_p) = \delta m\, c^2 \approx (10.53476 \cdot n_p + 6.01794 \cdot n_n)$ MeV

The regression coefficient of n_p (10,535) has a standard deviation σ of 0.135 and the regression coefficient of n_n (6.018) has a standard deviation σ of 0.095. The coefficient of determination R^2 indicates that 99% of the variation in binding energies is explained by the variations in their proton and neutron numbers. Relation 8.3 indicates that for every additional proton added to the nucleus, on average the binding energy increases by ≈10.5 MeV. For each neutron on average an increase of ≈6 MeV is found. Although for small nuclei, Relation 8.3 gives completely wrong binding energies, it is surprising that the ratio of the energy increase of a proton to that of a neutron amounts on average to about 7/4. The indicated ratio is given by

Relation 8.4 $10.53476/6.01794 \approx 1.750559 \approx 7/4$

If the average binding energy of ≈10.535 MeV is normalized with the characteristic energy $h_{bar}\, c\, L_2(\Lambda)^{-1}$ (≈ 12.640 MeV) from Table 8.2, then

Relation 8.5 $12.640/10.535 \approx 1.1998 \approx 6/5$

is obtained. It seems that the energy increase per proton is connected on average with an energy of $5/6\, L_2(\Lambda)^{-1}$. Because of Relations 8.4 and 8.5, the parameterization of the binding energy given by Relation 8.5 can be rewritten as

Relation 8.6 $B(n_n, n_p) = \delta m\, c^2 \approx 5/6 \cdot (n_p + 4/7 \cdot n_n) \cdot 12.640$ MeV

If all nucleons are removed from the nucleus and protons and neutrons are considered to be autonomous particles, on average binding energy is apparently released according to Relation 8.6. Which physical process is responsible for the energy difference to the experimentally measured value? Is the deviation from the average a manifestation of chaotic nuclear dynamics of interacting protons and neutrons, which, despite their interactions, approximately retain their identity? But, neither protons nor neutrons exist as autonomous particles in the nuclei.

In different nuclides, the mean binding energy per nucleon $B/(n_n + n_p)$ varies. It is lower for light nuclides than for heavy nuclides and is approximately constant for nuclides with atomic numbers greater than 20. The arithmetic mean of the binding energies per nucleon of all 2,932 isotopes compiled by Thayer Watkins [8.5] is ≈8.03 MeV. Interestingly, this corresponds to ≈12.640·(2/π) or ≈8.05 MeV, respectively. The standard deviation of a single value from the arithmetic mean value is ≈0.670 MeV and is therefore of the order of magnitude of $m_e c^2$. For more on this topic see Section 14.1.

Why can the binding energy of thousands of nuclei on average be parameterized by Relation 8.6? Is it a coincidence or an alternative view to the models of mainstream physics that almost exclusively attribute mass deficits to interactions of light quarks and massless gluons?

8.2.1 The light quark masses of baryonic matter in the reductionist view of the Standard Model

According to the Standard Model of particle physics, a proton consists of a down quark and two up quarks and a neutron of two down quarks and an up quark. The strong interaction between the quarks is mediated by massless particles called gluons. Free quarks cannot be observed and both quark varieties structurally resemble the electron. A proton and a neutron together therefore consist of three down quarks and three up quarks. The masses of quarks, which in baryons always occur in groups of three quarks distinguished by flavors, are free parameters of quantum chromodynamics. As a consequence, quantum chromodynamics cannot explain the masses of quarks. They must be determined by matching theory and experiment. The only argument for the "physical existence" of quarks was initially the observation that the inelastic scattering of electrons on nucleons is inhomogeneous. The proton and the neutron behave as if they consist of "hard" objects, on which electrons of high energy are elastically scattered.

The Particle Data Group (PDG), an international collaboration of particle physicists, compiles the mass of the up quark m_u at (2.5 + 0.6/−0.8) MeV/c^2, and the down quark is a little heavier and its mass m_d is given as (5 + 0.7/−0.9) MeV/c^2. Defining a mass for quarks is not easy and the values are therefore very controversial, because quarks exhibit confinement, which means that the quarks are not observed independently but only in combination with other quarks. Their values are therefore subject to regular changes. In any case, quarks cannot be considered as ordinary matter even though they have mass. The average quark mass of the six quarks in a neutron-proton pair is (3·2.5+3·5)/6, or ≈ 3.75 MeV/c^2, respectively. Thus, three quark masses in a proton or in a neutron correspond to ≈11.3 MeV/c^2. It is astonishing that this simple, naive calculation of the sum of the masses of three confined quarks yields $\approx h_{bar}\, c\, L_2(\Lambda)^{-1}$ per nucleon.

8.3 The universal gravitational interaction length and the MOdified Newtonian Dynamics model

Measurements of the rotation of galaxies or gas clouds show that the velocities outside the galaxy arms do not decrease as expected with increasing distance from the center of mass, but tend to be nearly the same. Similar to the solar system, where the outer planets move more slowly around the sun, objects in galaxies far from the center of mass should be falling in a Keplerian fashion. Since the classical law of gravitation does not provide an explanation for the flattening of rotational velocities, astronomers assume that in addition to the visible starlight and gas, dark matter exists in the outer region of galaxies, which is responsible for the increased velocities, and ensures that the outer stars, despite the centrifugal forces maintain their orbits. For to this day, no Keplerian decline of rotational curves has ever been observed, as predicted by Newton's laws. The suggested explanation for the discrepancy between the luminous mass and the dynamical mass of galaxies by means of hidden dark matter implies that Newton's laws of gravitation can be maintained unchanged, if the existence of such particles can be proved. Though despite intensive experimental research no dark matter particles have been found to date that are suitable for this purpose. To explain many experimental findings, however, dark matter is needed because no alternative can explain all phenomena.

An alternative to the dark matter paradigm in the form of a **MO**dified version of the **N**ewtonian **D**ynamics (MOND) was suggested by Mordehai Milgrom [8.6] in the year 1983. He recommended empirically changing the Newtonian law of motion in the limit of small accelerations so that for $a > a_0$ the standard expression $a = a_{Newton} = M\,G/R^2$ and for $a < a_0$ the relationship $a^2/a_0 = a_{Newton}$ is valid. The parameter a_0 is a constant of the dimension of an acceleration ($m\,s^{-2}$) and is considered to be a universal quantity whose cause is completely unclear and for whose calculation there is no theory to date. With the proposed modification various characteristics of galaxies can accurately be reproduced without the assumption of hidden mass. The freely selectable acceleration parameter a_0 must be determined by fitting the formulas of Newton's law of gravitation to observed rotation curves. K. G. Begeman and coworkers [8.7] obtained a value for a_0 of $(1.21 \pm 0.24)\cdot 10^{-10}\ m\,s^{-2}$ by radio observations in the 21 cm line of neutral hydrogen far beyond the optical disk.

Since both the dark matter paradigm and the model MOND are possible explanations for asymptotically flat rotation curves, Riccardo Scarpa and coauthors [8.8] investigated globular clusters. These objects, according to traditional doctrine, have little dark matter, and a distinction between the two models should be possible because they should obey Newton's laws for all accelerations due to lack of dark matter. The authors surprisingly found that in all globular clusters studied, the rotational curves at distances from the center, at which the gravitational accelerations reach the critical value a_0 of $(1.8 \pm 0.4) \cdot 10^{-10}$ m/s², nevertheless become asymptotically flat. Thus, MOND is well suited not only for galactic but also for subgalactic scales to phenomenologically describe observational data.

For the experimental determination of the critical acceleration parameter a_0, exact measurements of the baryonic masses and the asymptotically flat rotational velocities of the galaxies are needed. While the distance-independent, asymptotic rotation velocities are easy to obtain, measurements of the total masses of star-dominated spiral galaxies – such as our own Milky Way – have great uncertainties. To avoid this problem Stacy S. McGaugh [8.9] studied gas-rich galaxies, where the mass determination of stars plays a subordinate role, because the total mass is largely dominated by atomic gases, whose mass contribution can be more accurately determined. If a_0 was kept constant, an analysis of the data showed that all values statistically scatter around the average of $1.21 \cdot 10^{-10}$ m/s² determined by K. G. Begeman and coworkers [8.7] and that observational uncertainties were sufficient to explain the scatter. If a_0 was released as a parameter and the data of all 47 galaxies were reinvestigated, the value estimated for a_0 was $(1.24 \pm 0.14) \cdot 10^{-10}$ m/s².

Whatever attitude is taken to the model MOND, the enigmatic parameter a_0 is able to excellently describe flat rotation curves of hundreds of astronomical objects of different sizes not only qualitatively but also quantitatively. With MOND even details in the rotation curves can be reproduced from details in the distribution of baryonic masses. Although MOND is a great empirical success, in describing galaxy clusters where it is not possible to measure rotational curves, the model does not work and must assume additional dark matter like the Standard Model. Luminous matter alone is not adequate to establish the dynamics of interacting galaxies. Likewise, MOND is unable to provide explanations for gravitational lensing, structure formation, cosmology or the microwave background temperature.

The gravitational length l_{MOND} defined by

Definition 8.1 $l_{MOND} \equiv (G\, m_{grav} / a_0)^{1/2}$

shall be a contribution to the controversy about the model MOND, which accounts very well for the scale of galaxies, but does not tell much about the larger universe. With CODATA values and a Milgromian gravitational acceleration parameter a_0 of $\approx 1.1 \cdot 10^{-10}$ m/s², a value of ≈31 fm for l_{MOND} is obtained from Definition 8.1. Surprisingly, for the microscopic length scale l_{MOND},

Relation 8.7 $l_{MOND} \approx 2\, L_2(\Lambda) \approx 31$ fm

holds, which establishes a link of the universal acceleration constant a_0 to natural constants. The length scale $L_2(\Lambda)$ is obviously also a means to explain the parameter a_0 of the MOND model, that is, to shed light on enigmatic phenomena, which are difficult to explain using today's main stream models. Is Relation 8.7 a possible explanation for why the internal dynamics of galaxies do not obey the classical laws of Newton and Einstein at accelerations smaller than $\approx 1.1 \cdot 10^{-10}$ m/s²? Why is the proportionality factor $l_{MOND}/L_2(\Lambda)$ of the order of 2?

If the scales $L_2(\Lambda)$ and m_{grav} are the deeper, still lacking basis for the understanding of the phenomenological MOND formalism or the dark matter paradigm postulated by astronomers, it substantiates in an impressive way that astrophysics and microphysics are closely interwoven, and phenomena of the macrocosm can be traced back to parameters of the microcosm. Or is Relation 8.7 an irrelevant numerical coincidence for a phenomenological formalism that explains why stars at the edge of galaxies rotate faster than expected?

References

[8.1] D. C. Hoffman, F. O. Lawrence, J. L. Mewherter and F. M. Rourke. Detection of Plutonium-244 in nature. Nature 234 (1971) 132–134

[8.2] Ingo Sick. On the rms-radius of the proton. Physics Letters B 576 (2003) 62–67

[8.3] Peter G. Blunden and Ingo Sick. Proton radii and two-photon exchange. Physical Review C 72 (2005) 057601 [2 pages]

[8.4] Randolf Pohl, Aldo Antognini and Franz Kottmann. The size of the proton. Nature 466 (2010) 213–216

[8.5] Thayer Watkins. The enigma of the mass deficits of nuclei. http://www.sjsu.edu/faculty/watkins/massdeficits.htm, queried on October 19, 2013

[8.6] M. Milgrom. A modification of the Newtonian dynamics as a possible alternative to the hidden mass hypothesis. Astrophysical Journal 270 (1983) 365–370

[8.7] K. G. Begemann, A. H. Broeils and R. H. Sanders. Extended rotation curves of spiral galaxies:dark haloes and modified dynamics. Monthly Notices of the Royal Astronomical Society 249 (1991) 523–537

[8.8] R. Scarpa, G. Marconi and R. Gilmozzi. Globular clusters as a test for gravity in the weak acceleration regime. arXiv:astro-ph/0601581v1

[8.9] Stacy S. McGaugh. A novel test of the modified Newtonian dynamics with gas rich galaxies. Physical Review Letters 106 (2011) 121303 [4 pages]. arXiv:1102.3913v1 [astro-ph.CO]

9 The classical concept of the electrostatic field energy

I am inclined to suspect that the renormalization theory is something that will not survive in the future, and that the remarkable agreement between its results and experiments should be looked on as a fluke.
(*Paul Dirac*)[1]

The field concept is an elegant method for describing many-body systems. A field is a state of space without matter, but can be detected by effects on test objects. Likewise, a field is a carrier of energy. The concept of the physical field has proven to be very fruitful, since field theory allows a unified explanation of multiple physical processes. According to our knowledge, our cosmos as well as atoms consist of almost empty space and the observed masses are strongly localized. Based on this fact, matter particles are confined in a certain range of volume, and thus, because of the large number of particles present, only a statistical description by means of pressure, density or temperature is meaningful within this volume. Outside the volume, space is filled with electromagnetic fields. It makes sense to choose only one field type so that space gets a uniform structure.

9.1 The electrostatic pressure and the universal energy density

A finite volume can be considered as a charged sphere with a charge Q evenly distributed over its surface acting as the source of a field that alters the space around it. Such a sphere of radius r with a homogeneously charged shell generates an electric field E, which disappears in interior space and follows the law of Coulomb according to

Formula 9.1 $E = Q/(4\pi\varepsilon_0 R^2)$

in exterior space. The variable R represents the distance to the center of the sphere. If the sphere is at rest, Formula 9.1 can be used to calculate the energy density of the electrostatic field in outer space, which is dependent on the distance R. The calculation yields

Formula 9.2 energy density $(R) = (\varepsilon_0/2)\, E^2 = Q^2/(32\,\pi^2\,\varepsilon_0\,R^4)$

Since pressure (N/m^2) and energy density (J/m^3) have the same dimension, the energy density of Formula 9.2 is also called the electrostatic pressure or charge pressure.

A charged particle, such as the electron, can be considered as a perfectly conducting shell of radius r carrying a total charge q_e. But such a sphere would explode

1 The evolution of the physicist's picture of nature, Scientific American 208 (1963) 45 – 53.

https://doi.org/10.1515/9783110612387-009

because of the outward Coulomb pressure. In order to stabilize the sphere or the particle, the outward Coulomb pressure must be balanced by an inward pressure of the same amount.

A sphere in equilibrium cannot be a purely Maxwellian electromagnetic field implying a field-free empty vacuum state, but must additionally have a contracting property, which leads to an attractive force. Assuming that, for $Q = q_e$ and $R = r_{rho}$ the contracting property is caused by the virtual energy density rho c^2 defined in Chapter 6,

Definition 9.1 $(\varepsilon_0/2)\, E^2 = \text{rho}\, c^2 = q_e^2/(32\, \pi^2\, \varepsilon_0\, r_{rho}^4)$

is obtained. From this simple stability requirement, a critical shell radius r_{rho} for balance can be calculated. That is, at a certain critical distance r_{rho}, both repulsion and attraction are present. The calculation of the equilibrium radius from Definition 9.1 amounts to a value of r_{rho} of ≈123 nm.

9.1.1 The model of Hendrik Casimir for the calculation of the fine-structure constant

The classical electrodynamics is characterized by Maxwell's equations assuming a field-free, empty vacuum state. Without boundary conditions for the electric and magnetic fields, no electromagnetic forces are present. The Dutch physicist Hendrik Casimir proposed replacing the empty vacuum state with a background field of fluctuating electromagnetic radiation whose zero-point fluctuations should give rise to an attractive force to balance the Coulomb pressure. This would also open the way to calculate the fine-structure constant.

H. Casimir first clarified what happens when two conducting plates are pushed closer and closer together in vacuum and said that at small distances the influence of the vacuum energy should experimentally be detectable. Because certain modes of virtual oscillations are missing in the space between the metal plates, the quantum pressure of the exterior, where all modes can exist without restriction, should be greater. In vacuum, a force should therefore compress the plates that serve as an ideal mirror for virtual photons. Actually, it is amazing that summing up zero-point energies of infinite modes of virtual oscillations yields a finite result for the Casimir effect. The mathematical calculation is based on a difficult interplay of the infinite energies of interior space and exterior space, whose difference must yield a finite number, since a measurable result cannot depend on how large the intermediate stages are.

Zero-point fluctuations due to parallel plates give rise to an attractive force, but this result cannot be used for the calculation of the fine-structure constant because of the geometry. Therefore, T. H. Boyer [9.1] calculated the Casimir effect for a perfectly conducting spherical shell. He found that the Casimir effect is repulsive for spherical boundary conditions and not attractive, as originally calculated by H. Casimir for parallel metallic plates. For a charge-neutral, perfectly conducting hollow sphere with radius r, the electromagnetic Casimir energy, that is, the difference of the vacuum energy with and without geometrical boundary conditions, is given [9.2] by

Formula 9.3 $\varepsilon_{Casimir} = 0.046177... \cdot h_{bar}\, c\, r^{-1}$

If the Casimir energy given by Formula 9.3 is divided by the spherical volume $(4/3)\,\pi\, r^3$, the Casimir energy density results.

The result of T. H. Boyer was later confirmed many times by the use of other regularization and renormalization methods. Obviously, the Casimir effect of a spherical shell with a homogeneous charge distribution cannot serve as a stabilizing factor for the Coulomb pressure because it is repulsive, that is, the hope was shaken that zero-point fluctuations, although based on physical principles, can be used as a counter-force for space charge rejection. On the other hand, this finding aroused the hope of experimenters that due to the topological nature of the effect, both the sign and the distance dependence of the Casimir force could be specifically controlled by tailoring the shapes of the surfaces involved.

9.1.2 The experiment of U. Mohideen and Anushree Roy

U. Mohideen and Anushree Roy [9.3] measured the force between a Pd/Au-metallized polystyrene sphere attached to the cantilever of an atomic force microscope (AFM) and a metallized sapphire plate as a function of distance. A change in distance was achieved by moving the optically polished sapphire plate in steps of approximately 4 nm relative to the sphere with a diameter of ≈200 μm. The deflection of the cantilever tip was measured by means of a laser beam as a difference signal of a position-sensitive photodetector. The experiment was carried out at room temperature and at a pressure of 50 mTorr. U. Mohideen and Anushree Roy observed that the force curve, that is, the deflection of the AFM tip caused by the force between the sphere and the sapphire plate, noticeably changed at a distance of ≈100 nm, and markedly deviated from the linearity of the difference signal of the photodiode for greater distances.

Is the experiment of U. Mohideen and Anushree Roy an experimental manifestation of the length r_{rho}? Could the length r_{rho} also be the reason why in nanoelectromechanical systems' small-sized parts, which are less than 100 nm apart, spontaneously stick together and actuators are blocked? The common interpretation attributes these phenomena solely to electrostatic interactions, capillary effects or dispersion forces, because no universal length scale that fits the observations is available.

9.1.3 The experiment of H. B. Chan and coworkers

The quantum electrodynamical effect predicted by Casimir and the experimentally observed stickiness of nanoscale, electrically neutral structures motivated H. B. Chan and coworkers [9.4] to investigate the influence of this enigmatic force on the dynamic properties of flexible microstructures.

The micromechanical oscillator, used by H. B. Chan and coworkers in their investigations, consisted of a 3.5 μm thick, 500 μm × 500 μm polysilicon plate coated on the top with gold, which could freely rotate about two torsional rods (40 μm × 4 μm × 2 μm) on opposite edges. The ends of the torsional rods were connected on both sides via a pillar to the polysilicon substrate which was 2 μm lower than the flexible plate. The torsional mode of oscillation was excited by means of an excitation electrode (500 μm × 250 μm) which was located below the oscillator on one side of the torsional axis. On the other side of the torsional axis, also capacitively coupled to the top plate, a detection electrode (500 μm × 250 μm) was used to measure the induced oscillatory motion.

To investigate the effect of the Casimir force on the oscillator, a 200 μm diameter polystyrene sphere was placed close to one side of the oscillator. At a sphere distance of several micrometers, the oscillator behaved linearly at low excitation and the amplitude-frequency characteristic of the oscillator showed a symmetric resonance peak, that is, the Casimir force had a negligible effect on the oscillation. If the polystyrene sphere was brought closer to the oscillator, while maintaining the excitation strength, the resonant frequency shifted to lower frequencies. Moreover, the oscillation amplitude-frequency characteristic became asymmetric and exhibited hysteretic, highly nonlinear behavior at small oscillator-to-sphere distances. A stable operation of the oscillator seemed to be no longer possible at small distances due to the nonlinearity introduced by the Casimir force. The hysteresis was only reproducible if the measurement was started at the same plate-to-sphere distance.

In an alternative experiment, the oscillator was excited at a fixed frequency and the plate-to-sphere distance varied. The oscillation amplitude-distance characteristic showed pronounced hysteretic behavior with three regions. At an oscillator-to-sphere distance of ≈118 or ≈123 nm, respectively, the oscillation amplitude changed abruptly. Obviously, the oscillator acted as a sensor for the separation between the sphere and the oscillator surfaces with a pronounced memory effect.

There is no indication in the work of H. B. Chan and coworkers why the oscillation amplitude as a function of distance jumps at ≈118 or ≈123 nm, respectively. All experiments of the authors were performed at room temperature and at a pressure of less than 1 mTorr. It is interesting that an instability of the oscillation amplitude arose at a critical plate-to-sphere distance, which fits very well to r_{rho} of ≈123 nm, and thus the length r_{rho} given by Definition 9.1 appears to be experimentally confirmed very accurately.

Unfortunately, no experiments were found in the literature, which could provide further indications of the universality of the critical separations at ≈118 and ≈123 nm, respectively. Is it a coincidence that the memory effect of the micromechanical oscillator occurred at ≈123 nm and exactly matches r_{rho}? Does the experiment of H. B. Chan and coworkers allow an indirect determination of the vacuum energy density rho c^2?

9.2 The total field energy and the corresponding temperature of the different field types

If the energy density given by Formula 9.2 is integrated over the volume of the entire space from R = r to R = ∞, the totally stored field energy contained in the electrostatic field surrounding the charged spherical shell is obtained. The result is given by

Formula 9.4 $w_{\text{electrical field energy}} = w(r) = \varepsilon_0/2 \int E^2\, 4\pi R^2\, dR = Q^2/(8\pi\, \varepsilon_0\, r)$

For the calculation, the space was partitioned into concentric spherical shells of area $4\pi R^2$ and thickness dR. The parameter r denotes the radius of the homogeneously charged sphere. From Formula 9.4 it can be seen that for a simple point charge ($r \to 0$) an infinite amount of energy is stored in the field. However, this does not make any sense physically. It is therefore more appropriate to consider only energy contents of electric fields of homogeneously charged spheres whose radius r is finite. Otherwise, physical difficulties are predetermined. Indeed, in today's physics, such infinities are sometimes overcome by applying the mathematical trick of renormalization, so that the charge radius can even be reduced to zero without obtaining infinite values.

With Formula 9.4 and setting $Q = q_e$, field energies w(r) of different field types, that is energies of charged spherical shells with different radii r, can be calculated. This realizes the idea of the localization of an energy in a field. "Field temperatures" can also be assigned to electric field energies. Since the temperature is an intensive quantity, such a procedure is not a contradiction. The assignment is based on the assumption that energy and temperature are related and can be converted into each other by the Boltzmann constant k_B. Using various characteristic radii gives

Relation 9.1 $T(r = r_e/2) = w(r = r_e/2)/k_B \approx T(r \approx r_{\text{pion\_charged}}) \approx 5.9 \cdot 10^9$ K

Relation 9.2 $T(r = \lambda_e) = w(r = \lambda_e)/k_B \approx 3.44 \cdot 10^6$ K

Relation 9.3 $T(r = a_H) = w(r = a_H)/k_B \approx 1.58 \cdot 10^5$ K [≈ 13.6 eV]

Relation 9.4 $T(r = a_B) = w(r = a_B)/k_B \approx 3{,}321$ K

Relation 9.5 $T(r = r_{rho}) = w(r = r_{rho})/k_B \approx 67.79$ K [≈ 47.1 cm^{-1}]

Relation 9.6 $T(r = \Lambda) = w(r = \Lambda)/k_B \approx 1.929$ K

Relation 9.7 $T(r = L) = w(r = L)/k_B \approx 0.659$ K

For certain radii, field-specific identities can be derived. They are given by

Relation 9.8 $w(r = r_e/2) = q_e^2/(4\pi\, \varepsilon_0\, r_e) = h\, c/\lambda_e = m_e\, c^2$

Relation 9.9 $w(r = \lambda_{Lo}) = q_e^2/(8\pi\, \varepsilon_0\, \lambda_{Lo}) = h\, f_p/2$ [≈ 34.3 meV]

Relation 9.10 $w(r = a_B) = q_e^2/(8\pi\, \varepsilon_0\, a_B) = h\, c\, \Lambda^{-1} = E_B = m_0\, c^2$

Relation 9.11 $w(r = a_H) = q_e^2/(8\pi\, \varepsilon_0\, a_H) = 1/2 \cdot \alpha^2\, m_e\, c^2$ [≈ 13.6 eV]

9.2.1 The electron-positron reaction

As more and more heat is supplied to a system, the kinetic energy of the particles increases. However, the supply of energy is not possible without restrictions, since from a certain amount of energy particle-antiparticle pairs are spontaneously formed, which immediately radiate again. The released energy is used to heat the electromagnetic radiation field until thermal equilibrium between the creation and annihilation of electron and positron occurs. The electron-positron reaction can take place with a minimum photon energy of $m_e c^2$ (≈ 511 keV) per photon. This condition corresponds to a temperature of $\approx 5.9 \cdot 10^9$ K. Is $T(r = r_e/2)$ the critical temperature which can at most occur as a temperature of a heat bath in the observable universe? Is this temperature still a meaningful concept for describing a physical equilibrium state?

Radiation of the electron-positron annihilation was first discovered in the Milky Way in 1973. However, only the sensitive spectrometer on the ESA satellite INTEGRAL showed very clearly that the emission region of the galactic 511 keV annihilation line is centered in our Milky Way and has a spherically symmetric extension. The rest of the sky is dark in all other directions with respect to this radiation [9.5]. There is no evidence of a point-like source in addition to the diffuse emission from the galactic bulge region. The interpretation of these findings is unclear. What is certain is that the emission cannot be attributed to nucleosynthesis because otherwise the radiation would have to be present in the entire galactic plane.

It is noteworthy that the radius $r_e/2$ of the sphere, whose field energy corresponds to the rest energy of the electron, is approximately equal to reduced Compton length of the charged pion $r_{pion\_charged}$ calculated by means of its mass according to Definition 6.8. Using CODATA values for r_e and $r_{pion\_charged}$ gives

Relation 9.12 $r_e/2 \approx 1.409 \text{ fm} \approx r_{pion\_charged}$

If α is replaced by α_{geom} and m_e by m_{e_geom} of Relation 6.4, then $r_e/2$ is only a function of h and c and amounts to $\approx$1.413 fm. This value is surprisingly close to the value of the reduced Compton length of the charged pion of $\approx$1.414 fm and the already long-known, unexplained coincidence is improved. But the question what half the classical electron radius has to do with nuclear phenomena remains unanswered.

A numerical comparison reveals that

Relation 9.13 $r_e/2 \approx 2\pi\, r_{grav}$

holds. The radius $r_e/2$ of the sphere, whose field energy corresponds to the rest energy of the electron, is obviously similar to the Compton length of the virtual particle of mass m_{grav}. Using the definition of r_e, Definition 6.7 of r_{grav}, Relation 6.24 and $\alpha = \alpha_{geom}$

Relation 9.14 $r_e / (4\pi\, r_{grav}) = 2^{-11}\, \pi^{20/3} \approx 1.0069$

can be deduced. All occurring natural constants are canceled. The physics of all these coincidences is not clear.

The nucleon density of all nuclei except that of ^{4}He is fairly constant. The density of ^{4}He is abnormally high and about two times larger than the "standard density" of 0.17 nucleons per fm^3 of other nuclei. See also Section 8.1.1 for this topic. Setting the interaction radius to $r_e/2$, or $\approx r_{pion\_charged}$, or $\approx 2\pi\, r_{grav}$, a nucleon density of $(r_e/2)^{-3}$ or 0.36 nucleons per fm^3 results, which is indeed the density of ^{4}He experimentally observed. Which is the decisive parameter?

9.2.2 The primordial abundance of ^{4}He

Element abundances are very similar throughout the observable universe. Atomic hydrogen is the most abundant element and helium the second most common element. The percentage of all other elements is extremely small compared to hydrogen and helium. Assuming that free neutrons either disintegrate or become bound to ^{4}He, the relative abundance of ^{4}He is given by

Definition 9.2 $Y_{He} \equiv 2\eta/(\eta+1)$

where η represents the ratio of the number of neutrons n_n to the number of protons n_p in thermal equilibrium. For η

Ansatz 9.1 $\eta \equiv n_n/n_p = 7/4 \cdot \exp\{-(m_n - m_p)\, c^2/(k_B T)\}$

shall apply. The parameter m_n is the mass of the neutron and m_p is that of the proton. The exponential function originates from the Boltzmann distribution, and the prefactor is the phenomenological ratio 7/4, which also showed up in connection with the number of protons and the number of neutrons in Relation 8.4. Ansatz 9.1 states that in thermal equilibrium protons and neutrons are transformed into each other by electrons and neutrinos, and the transformation process depends on the rest energy or rest mass of the heavy particles involved.

If the temperature $T(r = r_e/2)$ of $\approx 5.9 \cdot 10^9$ K of Relation 9.1 is used for the temperature T in Ansatz 9.1, a value of ≈ 0.139 is obtained for the ratio of the number of neutrons to the number of protons η. Using η in Definition 9.2, a value of $\approx$**0.244** thus results for the primordial helium abundance Y_{He}. Using Relation 9.8 it is easy to show that

Relation 9.15 $\eta = 7/4 \cdot \exp\{-(m_n - m_p)/m_e\}$

is equivalent to Ansatz 9.1. Experimentally, ^{4}He must be searched for as far away as possible, so that the measurement is not contaminated by stellar ^{4}He. Emission lines of optical recombination lines of ^{4}He in metal-deficient extragalactic H II regions are usually used to measure the primordial abundance of ^{4}He. The observed [9.6] primordial helium abundance of 45 H II regions amounts to **0.244** ± **0.002**, which is in excellent accord with the value calculated according to Ansatz 9.1.

9.2.3 Remarks on $T(r = r_{rho})$

For the case $r = r_{rho}$, the stored field energy $w(r = r_{rho})$ can easily be converted into a volume, if $Q = q_e$ and $r = r_{rho}$ from Definition 9.1 are substituted into Formula 9.4. In addition, the expression $rho \cdot c^2$ must be replaced by $(8\pi)^{-1}$ J m^{-3} according to Relation 6.1. The result is

Relation 9.16 $w(r = r_{rho}) = 2^{-1}\, r_{rho}^3$

Although there is an energy on the left side and a volume on the right side of Relation 9.16, the unit m^3 has to be replaced by the unit J in this particular case. Dividing the term $2^{-1}\, r_{rho}^3$ by c^2 then gives the unit kg, which allows a numerical comparison with the mass m_0. Using CODATA values

Relation 9.17 $2^{-1}\, r_{rho}^3/c^2 \approx 48.995^{-1} \cdot m_0 \approx 7^{-2}\, m_0$

ensues. The rest energy m_0 is thus larger by a factor of $\approx 7^2$ than the stored field energy of the sphere of radius r_{rho}. It is interesting that approximately the number seven pops up again. Geometrizing Relation 9.17, that is setting $\alpha = \alpha_{geom}$ and $m_e = m_{e\_geom}$ according to Relation 6.4,

Relation 9.18 $2^{-1}\, r_{rho}^3 / (m_0 c^2) = 2^{-3}\, \pi^{-19/12} \approx 49.005^{-1}$

follows. The conversion of the field energy $w(r = r_{rho})$ into the corresponding temperature can be accomplished by dividing Relation 9.16 with the Boltzmann constant k_B. A comparison of the temperature thus calculated, using CODATA values for the natural constants, with the reference temperature τ leads to

Relation 9.19 $2^{-1}\, r_{rho}^3/k_B \approx 0.6999{\cdot}\alpha^{1/2}\,\tau \approx (1/2)\,(7/5)\,\alpha^{1/2}\,\tau$

or

Relation 9.20 $T(r = r_{rho}) \approx (1/2)\,(7/5)\,\alpha^{1/2}\,\tau \approx 67.79\ \mathrm{K}$

Using the definitions for $w(r = r_{rho})$ and E_τ

Relation 9.21 $w(r = r_{rho}) \,/\, (E_\tau \alpha^{1/2}) = c^{-5/36}\, h^{1/12}\, m_e^{-1/9}\, \alpha^{-1/12}\, 2^{11/12}\, \pi^{-1/108}$

can be deduced after some calculation. The power product $c^{-5/36}\, h^{1/12}\, m_e^{-1/9}$ occurring in Relation 9.21 can be written as

Relation 9.22 $c^{-5/36}\, h^{1/12}\, m_e^{-1/9} = \{c^{5/18}\, h^{-1/6}\, m_e^{2/9}\}^{-1/2} = \{2^{5/6}\, \pi^{52/27}\}^{-1/2}$

because of Relation 6.3. Using Relation 9.22, and $\alpha_{geom}^{-1} = 2^6\, \pi^{2/3}$, the geometrized Relation 9.21 may be written as

Relation 9.23 $w(r = r_{rho})/(E_\tau \alpha^{1/2}) = 2\,\pi^{-11/12} \approx 0.7003 \approx 1/2 \cdot 7/5$

9.2.4 The spatial hysteresis of the microelectromechanical actuator

H. B. Chan and coworkers [9.4] observed in their experiment that the oscillation amplitude depended strongly on the distance characteristics and a stable and reproducible hysteresis behavior occurred, when a metallic sphere was brought close to the micro-oscillator. This experiment is described in detail in Section 9.1. The amplitude-distance characteristic has not only a discontinuity at ≈123 nm, which was assigned to the characteristic length r_{rho} in Section 9.1, but also exhibits a discontinuity at ≈118 nm. Forming the difference between these two discontinuities results in a value of ≈5 nm, which approximately corresponds to $2a_B$. It appears that the width of the spatial hysteresis is an approximate measure of the double Bohr radius of the virtual electron-hole pair introduced in Chapter 2, and, therefore, the micro-oscillator acts as a sensor not only for the length r_{rho} but also for the length a_B.

Relation 9.17 establishes a direct connection between the field energy $w(r = r_{rho})$ and the mass quantum m_0. From this, the ratio of the characteristic length r_{rho} to the length a_B can be derived by simple algebraic manipulations. It gives

Relation 9.24 $r_{rho}/a_B = (m_0c^2)/w(r = r_{rho}) \approx 7^2$

However, an explanation for the observed spatial hysteresis cannot be derived from Relation 9.24.

9.2.5 The photosphere of the sun or the energy scale of stars

The photosphere is the part of the sun from which almost all visual light and heat energy of the sun emerges. The effective mean temperature is about 5,778 K [9.7]. This temperature corresponds to the temperature of a blackbody of the same size, which emits the same total radiant power or luminosity. The photosphere is about 400 km deep, which is a small value compared to the sun's radius. The photosphere has a temperature of ≈6,600 K at the bottom boundary, and ≈4,400 K at the top boundary [9.7]. The photosphere's minimum temperature at the top boundary is followed by the chromosphere, in which the temperature rises again to $\approx 3{\cdot}10^4$ K [9.7]. Later, the chromosphere follows the corona, the outermost part of the sun's atmosphere, with a second increase in temperature of up to $\approx 2{\cdot}10^6$ K. Then the temperature drops continuously to the value of the universe of ≈3 K. The temperature data make it clear that the different layers of the sun are not in thermodynamic equilibrium, but stationary equilibria must be present if the concept of temperature makes sense.

If the temperatures of the sun's photosphere are compared with the temperature scale $T(r = a_B)$, that is, with $k_B^{-1}\, m_0\, c^2$, it is surprising that in units of $T(r = a_B)$ simple fractions visualized by

Relation 9.25 $5{,}780\ \mathrm{K}/T(r = a_B) \approx 1.74 \approx 7/4$

Relation 9.26 $4{,}400\ \mathrm{K}/T(r = a_B) \approx 1.32 \approx 4/3$

Relation 9.27 $6{,}600\ \mathrm{K}/T(r = a_B) \approx 1.99 \approx 2$

emerge. Could it be that m_0c^2 is the discrete energy scale of the sun's photosphere? Could this apply to other stars as well? Are stars in stationary states, that is, when they are in a stable equilibrium, "macroquantized" collective systems? Are effective temperatures of photospheres of stars in units of $T(r = a_B)$ simple fractions that can be formed from small natural numbers? Which numbers are allowed?

Experimentally, these questions are not easy to clarify, because the effective temperature itself is not directly observable. It must be calculated from the state variables luminosity and radius, or is based on star models. Despite this problem, Table 9.1 shows effective temperatures of a selection of stars from the literature and their normalization by the energy scale m_0c^2. The interpretation of the table is left to the reader. It is clear that many stars can be found in the literature, whose effective temperatures in units of m_0c^2 or $T(r = a_B)$ do not yield simple fractions.

$T(r = a_B)$ can be considered as the average thermal energy (≈3,321 K), which corresponds to the electrostatic interaction energy of an electron and a positron at a distance of $2a_B$ (≈ 5 nm). At mean particle distances, which are large compared to this critical distance or Landau length, there is a plasma that acts like an ideal gas, because the electromagnetic interactions can be neglected.

Table 9.1: Effective temperatures of some stars

Object	T_{eff} (K)	$T_{eff}/T(r = a_B)$		Reference
Sirius A	9,900	≈2.98	≈ 3	[9.8]
α-Centauri A	5,810	≈1.75	≈ 7/4	[9.9]
α-Centauri B	5,260	≈1.58	≈ 8/5	[9.9]
Procyon A	6,550	≈1.97	≈ 2	[9.10]
Procyon B	7,740	≈2.33	≈ 7/3	[9.11]
α-Draconis	9,988	≈2.99	≈ 3	[9.12]
τ-Herculis	15,011	≈4.52	≈ 9/2	[9.12]
γ-Lyrae	9,612	≈2.89		[9.12]
HR 7926	13,306	≈4.01	≈ 4	[9.12]
Arcturus	4,290	≈1.29	≈ 9/7	[9.13]
Capella A	4,940	≈ 1.49	≈ 3/2	[9.14]
Polaris	5,984	≈ 1.80	≈ 9/5	[9.15]

Note: The effective temperatures for α-Draconis, τ-Herculis, γ-Lyrae and HR 7926 are the means of the effective temperatures of two published works. The effective temperature of Polaris is an averaged value of 23 measurements during the years 1944–2004.

9.2.6 The ^{7}Li abundance of stars

If the ^{7}Li abundance [9.16] in the most metal-poor stars is plotted against their effective temperature, the abundance increases continuously to about 5,500 K. Above ≈5,500 K, the abundances remain fairly constant. Interestingly, this experimental observation scales quite well with the energy scale $T(r = a_B)$ in the sense previously discussed, as 5,500 K in units of $T(r = a_B)$ amounts to ≈1.66 or ≈5/3, respectively.

9.2.7 The solar corona

The solar corona is the spectacular outer atmosphere of the sun, which becomes visible as a halo during a total solar eclipse. Thermal Doppler widths of the coronal lines gave, as the most probable temperature [9.17] of the inner solar corona, a temperature of (2.4 ± 0.15) MK. If

Ansatz 9.2 $<\varepsilon_{kin}> = w\{r = \lambda_e = h/(m_e c)\} = (3/2)\cdot k_B T_{corona}$

is used for the mean corona temperature T_{corona}, a value of ≈**2.3·10⁶** K results, which is very close to the most probable temperature determined from Doppler widths.

J. W. Brosius and coworkers [9.18] derived an average corona temperature of (2.5 ± 0.1) MK from the emission lines of four different stages of ionization of iron. Interestingly, this value is also close to the value obtained from Ansatz 9.2. The reason for the high corona temperature is poorly understood despite hundreds of theoretical models, and baffled scientists for decades. The only consensus is that the x-ray emitting corona must have a permanent heating in order to maintain the observed steep temperature gradient between the chromosphere and the corona against thermal conduction. Is the field energy $w(r = \lambda_e)$ the energy source of the solar corona, which replenishes what is lost by the outward flux of high energy particles and the thermal conduction into and out of the sun?

The solar surface is not only crossed by sunspots, but it is also strewn with tiny, x-ray bright points (XBPs) that are so small that they can only be partially resolved with today's telescopes. R. Kariyappa and coworkers [9.19] found, by analyzing data from the x-ray telescope of the Japanese satellite Hinode, that the average temperatures of 14 XBPs ranged from 1.1 to 3.4 MK during a 7 h long time sequence, revealing that the heating rate is highly variable on this timescale. The mean value of 1.1 and 3.4 MK amounts to 2.25 MK, which is quite close to the value of ≈2.3 MK according to Ansatz 9.2. Is, as stated above, the field energy $w(r = \lambda_e)$ indeed acting as the energy source that heats up the solar corona in the form of hot bright points that randomly appear and disappear?

9.2.8 Grote Reber

The astronomer Grote Reber was a pioneer of radio astronomy. He was one of the first who investigated hectometer radio waves reaching the earth from the Milky Way. For such waves, the ionosphere acts as a two-sided mirror and the observation of hectometer radio waves can therefore only be accomplished through a hole in the ionosphere. Ionospheric holes, where little absorption of the hectometer waves by the plasma occurs, are very rare. This fact severely restricts the experimental possibilities and makes observations only possible at certain times and in places of low ion density and low solar activity. Grote Reber observed at a wavelength of 144 m (≈2.1 MHz) a bright galactic emission continuum in the southern sky of Tasmania viewing to the center of the Milky Way.

The radiation intensity received by a radio telescope can be converted into an apparent temperature. This temperature, according to the Rayleigh–Jeans law, corresponds to the temperature of a blackbody emitter, which at the observed wavelength (frequency) emits the same intensity of radiation as the source observed by the telescope. Grote Reber calculated [9.20] for the center of the Milky Way an apparent temperature of $\approx 3.5 \cdot 10^6$ K.

In the year 1968, the magnitude of the bright background was completely unexpected and no theory could explain or predict the experimental finding of Grote Reber. The source of the background, which seemed to be an all-pervading bright fog extending to infinity, must be outside the Milky Way and substantially isotropic. What is the source of the background measured by Grote Reber? May the electron be considered as a transducer of intergalactic light to radio energy? Interestingly, the apparent temperature, calculated from the measured radio power and the geometry of the antenna, corresponds to the field temperature $T(r = \lambda_e)$ of $\approx 3.44 \cdot 10^6$ K. Is this a numerical coincidence or after all a hint of a deep connection?

9.3 Raman lines of diamond and graphite: Are they based on interactions with fields?

Raman and infrared spectra are mostly dominated by the influence of short-range forces, that is, the atoms of the next environment. The spectra of amorphous and crystalline materials with similar short-range order are comparable for this reason. But, as per the below-mentioned examples, it is not always the case that spectra invariably refer to short-range properties of isolated atomic clusters or molecules.

9.3.1 Single-crystal sp³-hybridized diamond

The Raman spectrum [9.21] of a single-crystal, sp^3-hybridized diamond is dominated by a single, sharp line at ~1,332 cm^{-1}, which corresponds to the vibrations of the two interpenetrating cubic sublattices. The full width at half maximum of this line is very small and amounts to $\approx 1\ cm^{-1}$. Interestingly,

Ansatz 9.3 $\varepsilon_{diamond} = 7/4 \cdot E_T - w(r = r_{rho})$

gives a value of $\approx 1{,}332.1\ cm^{-1}$, if CODATA values for the natural constants are used to calculate the energies E_T and $w(r = r_{rho})$. Converting energy in joules to wave numbers is achieved by dividing energy in joules by the conversion factor $100 \cdot hc$. Why can a material property, that is, the first-order Raman line of diamond, be calculated solely by natural constants using a simple combination of E_T, $w(r = r_{rho})$ and the fraction 7/4? Is the phenomenological Ansatz 9.3 a coincidence? On its own, this may be true, but why can similar combinations with E_T and $w(r = r_{rho})$ be found for graphite?

9.3.2 Highly oriented pyrolytic sp²-hybridized graphite

In contrast to diamond, highly oriented pyrolytic graphite (HOPG) behaves like a pure metal, is reflective and electrically conductive and very brittle. The Raman spectrum of HOPG is more complicated than that of diamond and exceptional in several respects. In the case of large crystals, in addition to the first-order G line $\varepsilon_{graphite\_G\text{-}band}$ at ≈1,577 cm^{-1}, the spectrum also shows second-order lines whose first-order partners only exist in the case of crystalline disorder.

The first-order G line can, as outlined in Section 5.3.4, be approximated by

Ansatz 9.4 $\varepsilon_{graphite\_G\text{-}band} = 2E_T$

A very sharp second-order line, the so-called 2D′ line $\varepsilon_{graphite\_2D'\text{-}band}$, does not appear at double the value of ≈1,577 cm^{-1} as might be expected due to the order, but at ≈3,248 cm^{-1} [5.11], or at ≈3,247 cm^{-1} [5.12], respectively. The dependence of the anomalous 2D′ line on the excitation wavelength is very small. Interestingly,

Ansatz 9.5 $\varepsilon_{graphite\_2D'\text{-}band} = 4E_T + 2w(r = r_{rho})$

gives a value for the 2D′ line at ≈3,246.7 cm^{-1}, if CODATA values for the natural constants are used. Combining E_T, $w(r = r_{rho})$ and simple numbers leads again to a remarkable agreement with the experiment.

Besides the 2D′ line at ≈3,246 cm^{-1}, the Raman spectrum of HOPG shows an additional, strong 2D line $\varepsilon_{graphite\_2D\text{-}band}$ with a shoulder at ≈2,710 cm^{-1} [5.11]. The corresponding D line at ≈1,355 cm^{-1} is completely absent for large single crystals. Only polycrystalline graphite samples always show the first-order D line at ≈1,355 cm^{-1}. Even in nanocrystalline materials, this feature still remains quite sharp. It shifts, however, slightly to ≈1,360 cm^{-1}. If the energy of the 2D transition is given by

Ansatz 9.6 $\varepsilon_{graphite\_2D\text{-}band} = 7/2 \cdot E_T - w(r = r_{rho})$

the 2D mode appears at ≈2,711.3 cm^{-1} yielding a corresponding first-order D mode at ≈1,355.7 cm^{-1}. Both calculated positions are in good agreement with observations. It should be noted, however, that both the 2D line and the D line represent doublets whose positions may depend on the excitation wavelength.

9.3.3 Disordered sp²-hybridized graphite

The implantation of ^{12}C ions can artificially create disorder in HOPG. Due to disorder, the D′ line $\varepsilon_{graphite\_D'\text{-}band}$ appears in the first-order spectrum at ≈1,623 cm^{-1} [9.23]. The feature is mostly seen as a shoulder of the G line and depends only weakly on the laser excitation wavelength. Likewise, the D line and the 2D line appear as doublets. In contrast to the D′ line, the positions of the D and 2D lines depend on the laser excitation wavelength and shift approximately linearly with higher laser excitation energy to higher wave numbers. A dependence of the intensities and the widths of the D and the 2D modes are also observed. The anomalous, disorder-induced D doublet is observed in all sp^2-hybridized carbon materials and appears to be universal. Because of its complicated behavior, it is difficult to make quantitative statements about the D and 2D lines. Since the D′ line represents a single band with no dependence on the excitation wavelength, it is easier to assign. Using Ansatz 9.5, a value of ≈1,623.4 cm^{-1} is obtained for the D′ line, which agrees very well with the value of ≈1,623 cm^{-1} observed by Ping-Heng Tan and coworkers [9.23].

9.3.4 Splitting of the G line of graphite by aromatic adsorbates

The aromatic molecule tetrasodium 1,3,6,8-pyrenetetrasulfonic acid (TPA) can effectively exfoliate graphite into graphene monolayers with the aid of sonication in aqueous solutions. Such peeled-off, two-dimensional carbon crystals are plane, randomly bounded aromatic systems with many resonant π bonds. Graphene monolayers are very hard and have interesting physical properties even at room temperature. It is also possible to sandwich a monolayer of graphene between aromatic molecules, which can bind to the basal plane of graphene via π–π interactions, and dispense onto a substrate. Such graphene composites forming well-ordered structures by self-assembly can be characterized by Raman spectroscopy.

Xiaochen Dong and coworkers [9.22] found that a monolayer of graphene sandwiched between TPA and dispensed onto a SiO_2 substrate exhibited a laser-frequency-independent G band splitting of ≈24 cm^{-1}, which is otherwise only observed in carbon nanotubes. A similar splitting of the G line into G^- (≈1,567 cm^{-1}) and G^+ (≈1,590 cm^{-1}) was observed in graphene monolayers dispersed by other adsorbates, such as pyrene, anthracene or naphthalene. No graphene monolayers were found in suspension using monoaromatic rings, such as benzene or toluene. The center of the G^- and G^+ lines is located at ≈1,578 cm^{-1} for all investigated aromatics and agrees very well with $2E_T$. Interestingly, the splitting caused by the adsorption of aromatic molecules amounts to $w(r = r_{rho})/2$, or ≈ 23.5 cm^{-1}, respectively.

9.3.5 The dependence of certain Raman lines on the excitation energy

Several authors investigated the dependence of the disorder-induced Raman D or 2D band on the incident laser energy. Till now no satisfactory model is known to explain the physical origin of the observed dispersion of the D or 2D band with laser energy.

M. J. Matthews and coworkers [9.24] plotted the D band frequency ω as a function of the laser energy ε_L using data reported by Yan Wang and coauthors [5.12], data of polyparaphenylene and data of crystalline graphite. The ω versus ε_L data could be well fit by straight lines with the same slopes $\Delta\omega/\Delta\varepsilon_L$ of $\approx$51 cm^{-1} eV^{-1}. A linear regression for the second-order 2D band also showed the same slopes for all data, which were approximately two times the experimental slope for the D band.

Istvan Pocsik and coworkers [9.25] also studied the dependence of the D line on laser excitation energy in the range from 1.16 to 4.3 eV. They used polycrystalline POCO® graphite, which is easily available on the market, for their experiments. The positions of the G and D lines were determined by fitting the spectra to a single Lorentzian line. From the dependence of the line position on the excitation energy, they obtained for the slope $\Delta\omega/\Delta\varepsilon_L$ a value of $\approx$50 cm^{-1} eV^{-1}. Whereas the D band showed an apparent linear variation of its position with laser energy, the G line revealed virtually no dependence on the excitation frequency.

The dependence of the D and 2D bands in the Raman spectrum on the laser excitation energy can also be observed in TPA-dispersed graphene monolayers. Xiaochen Dong and coworkers [9.22] obtained for $\Delta\omega/\Delta\varepsilon_L$ a value of $\approx$51 cm^{-1} eV^{-1} for the D line using three different laser excitation energies to evaluate the slope. Accordingly, the frequency-energy dispersion $\Delta\omega/\Delta\varepsilon_L$ for a monolayer of graphene, sandwiched between two parallel films of TPA molecules, is of the same order of magnitude as that for disordered graphite.

The frequency-energy dispersion $\Delta\omega/\Delta\varepsilon$ can be understood as an energy or frequency ratio and is therefore a dimensionless parameter. If numerator and denominator are converted into the same units, then

Ansatz 9.7 $$\Delta\omega/\Delta\varepsilon_L \approx 51\ \mathrm{cm}^{-1}\ \mathrm{eV}^{-1} \approx 51\cdot c\cdot h\cdot 10^{2}\cdot q_e^{-1} \approx 0.866\cdot\alpha \approx \sin(120°)\cdot\alpha$$

results using CODATA values for the natural constants. It is astonishing that $\Delta\omega/\Delta\varepsilon$ is on the order of the Sommerfeld constant α. It certainly needs more observational data to clarify if the proportionality factor in Ansatz 9.7 is indeed sin(120°). In any case, the universal dispersive behavior of sp^2-hybridized carbon can adequately be described by Ansatz 9.7 based on available data. It cannot be denied that a numerical factor sin(120°) in sp^2-hybridized carbon could be important due to geometrical reasons.[2]

---

2 The graphite lattice consists of parallel planes of C atoms whose bonds in the plane form a hexagonal structure. The planar bond angle to the next three C atoms is 120°.

9.4 The Boson peak and the thermal conductivity plateau in amorphous materials

In the case of glasses and crystals, elastic waves with large wavelengths can be very well described with the concept of phonons, that is, by the theory of Debye, as long as the wavelengths are so large that the medium can be regarded as homogeneous. At high frequencies, however, amorphous solids and crystals differ considerably, since disordered systems such as glasses have additional modes in addition to the Debye phonons. This manifests itself in the fact that the vibrational density of states (VDOS) shows an excess over the Debye density of states, which can be calculated from sound velocities. A distinct peak is often not seen in the VDOS itself but appears only in the reduced VDOS after normalizing by the squared frequency ω^2, which reveals the excess over the Debye model prediction which is proportional to ω^2. This excess appears as a broad band in the terahertz region and is called the Boson peak because its intensity changes with temperature as the Bose factor. The terahertz region is a frequency range where the Debye model for crystals still works very well, but for glasses, differences seem to become apparent.

Experimental evidence for the anomalous behavior of disordered materials compared to their crystalline counterparts comes mainly from inelastic neutron [9.26] and light scattering experiments, whose spectra show a broad, asymmetric band with almost universal shape at low energies. The broad Raman band experimentally observed in no way conforms to the sharp lines that are common in crystals at other energies. The maximum of the low energy band, which is completely absent in near-perfect crystals, is not always exactly at the same energy, but ranges from 2 meV ($\approx$16 cm^{-1}) to 8 meV ($\approx$64 cm^{-1}). The weak dependence of the Boson peak ε_{BP} on material properties is due to system-specific effects such as pressure, temperature and composition. Although the Boson peak in spectra can inaccurately be localized, it is believed that it is a universal feature and the key to understanding glass-forming materials and liquids. The nature of the additional modes in disordered materials is not understood and is therefore still the subject of intense and controversial debate in the literature. Many models have been suggested to explain the Boson peak, but a comprehensive understanding has proved elusive, because all scenarios cannot be distinguished from each other experimentally. If, according to today's opinion, the Boson energy ε_{BP} is a universal quantity, it must be calculable by means of natural constants. Using

Ansatz 9.8 $$\varepsilon_{BP} \approx w(r = r_{rho}) \approx 5.8 \text{ meV } (\approx 47 \text{ cm}^{-1} \text{ or } \approx 1.4 \text{ THz})$$

a universal value for ε_{BP} within the experimentally observed bandwidth is obtained, which is only dependent on natural constants, and therefore could serve as an explanation for the Boson peak.

Sound waves are plane waves with a defined wave vector k_{ph} linked to the angular frequency ω via the dispersion relation $\omega = 2\pi f = v_{sound}\, k_{ph}$. These acoustic waves propagate up to relatively high frequencies with a phase velocity v_{sound} and have a finite mean free path l_{ph}, in spite of the disorder. Beyond a certain crossover frequency one expects [9.26] that acoustic waves can no longer propagate freely, are strongly scattered and lose their wave character, if the average scatter length l_{ph} is of the same order of magnitude as the wave vector k_{ph}. This leads to the condition $l_{ph}\, k_{ph} \approx 1$, from which the expression $l_{ph} \approx v_{sound}\,(2\pi\, f_{BP})^{-1}$ for the mean free path can be derived, which relates the Boson peak frequency f_{BP} to the length scale l_{ph}. Quartz glass has the most pronounced Boson peak of all glasses and has a longitudinal sound velocity v_{sound} of ≈5,900 m s^{-1}. Using $l_{ph} \approx v_{sound}\,(2\pi\, f_{BP})^{-1}$, and $f_{BP} \approx 1.4$ THz from Ansatz 9.8, gives a characteristic length scale l_{ph} of ≈6.65 Å, which appears to be rather large, but is strikingly close to λ_{MI} given by Definition 4.1.

9.4.1 The plateau in the thermal conductivity of amorphous materials

The thermal conductivity κ in glasses is orders of magnitude smaller than that in crystals, practically identical for all vitreous materials and independent of chemical composition. At low temperatures the thermal conductivity of amorphous solids varies approximately at T^2 at temperatures up to ≈5 K. Between ≈5 and ≈15 K, κ remains almost constant and displays a nearly temperature-independent plateau. Above ≈15 K, the thermal conductivity slowly increases again.[3] The heat transport in the plateau is almost independent of the material and shows a “universal” thermal conductivity of ≈$10^{-3}\cdot$W$\cdot$cm$^{-1}\cdot$K^{-1} for almost all glasses [9.27]. Although the flattening of the thermal conductivity of glass-forming substances always occurs at about the same temperature $T_{plateau}$ of ≈10 K, the position of the plateau and the value of the thermal conductivity in the plateau $\kappa_{plateau}$ are superimposed by material properties, similar to the Boson peak. The close agreement of the coherence length l_{ph} of ≈6.65 Å with the critical electron wavelength λ_{MI} of the metal-insulator transition suggests the idea to correlate the electrical resistance ρ_{MI} and the thermal conductivity $\kappa_{plateau}$, in analogy to the rule of Wiedemann and Franz, according to

Ansatz 9.9 $\kappa_{plateau}\, \rho_{MI} = (7/2\, k_B/q_e)^2\, T_{plateau}$

Setting $\rho_{MI} \approx 981$ μΩ cm and $T_{plateau} \approx 10$ K, a value of ≈10^{-3} W cm^{-1} K^{-1} is obtained for the thermal conductivity $\kappa_{plateau}$, which is in good agreement with the experimentally observed value. What is the reason behind the experimental fact that $T_{plateau}$ amounts to ≈10 K? Setting $(7/2)\, k_B T_{plateau} = w(r = r_{rho})/2$, then $T_{plateau}$ yields ≈10 K. In this context, see also Ansatz 9.6.

3 Thermal conductivities of crystals at low temperatures usually follow a T^3 law as a function of temperature.

9.5 Rydberg atoms: the transition from quantum mechanics to classical physics

Highly excited atoms are often called Rydberg atoms. In such atoms, the outermost electron is in a state with a very high quantum number and moves almost classically around the nucleus shielded by other electrons. Rydberg atoms are extremely sensitive to their environment and show long-range interactions with a wealth of exotic properties. These highly excited atoms close to the ionization limit live for a remarkable long time, are invisible and virtually mark the threshold between classical and quantum mechanical physics. Since the binding energy is low in a state of high electronic excitation, Rydberg atoms radiate in the microwave range. Due to the fact that the electron is located far outside the compact ionic core made up of the nucleus and the inner electrons, the properties of a Rydberg atom are very similar to those of the hydrogen atom, and the allowed energies of the electron can be modeled analogously to the hydrogen atom with the Rydberg energy R_∞ and an effective quantum number n_{eff}, but which is no longer an integer as in the Bohr model. The allowed energies ε of the electron are given by

Formula 9.5 $$\varepsilon = R_\infty \, n_{eff}^{-2} = \varepsilon_{atomic} \, n_{eff}^{-2}$$

The quantum number n_{eff} designates the energy level of the electron and ε_{atomic} denotes the atomic energy scale given by the expression $2^{-1} \alpha^2 m_e c^2$. According to Bohr's correspondence principle, systems with high quantum numbers behave like classical systems, and according to this principle a transition should take place where the electron behaves like a classical oscillator, which radiates electromagnetic energy with a frequency equal to the orbital frequency. In the limit of high quantum numbers, that is, for the transition to classical electrodynamics, the binding energy shall heuristically be equated with the field energy $w(r = L)$ according to

Ansatz 9.10 $$\varepsilon_{atomic} \, n_{eff}^{-2} = w(r = L)$$

Replacing the field energy by Formula 9.4 with $Q = q_e$ and $r = L$, a relationship between the ratio L/a_H and the effective quantum number n_{eff} can be derived. It is given by

Relation 9.28 $$L/a_H = n_{eff}^2 = c^{10/9} \, h^{-2/3} \, m_e^{8/9} \, \alpha^{1/6} \, 2^{2/3} \, \pi^{38/27}$$

As can be seen from Relation 9.28, the ratio L/a_H depends only on natural constants. By using CODATA values, a critical value of ≈489.4 can be worked out for the largest achievable quantum number n_{eff}.

Setting $\alpha = \alpha_{geom}$ and $m_e = m_{e\_geom}$ according to Relation 6.4, n_{eff} squared or L/a_H, respectively, is given by

Relation 9.29 $$n_{eff}^2 = 2^3 \, \pi^9 = L/a_H$$

which gives a critical value of ≈488.3 for the maximum quantum number n_{eff}. The corresponding Bohr radius is the length L, since for all Rydberg atoms the mean orbital radius is given by $a_H\ n_{eff}^2$. Rydberg atoms in states with n_{eff} as large as 350 have been detected in space by radio astronomers.

References

[9.1] Timothy H. Boyer. Quantum electromagnetic zero-point energy of a conducting spherical shell and the Casimir model for a charged particle. Physical Review 174 (1968) 1764–1776

[9.2] K. A. Milton. The Casimir effect. World Scientific, 2001, New Jersey

[9.3] U. Mohideen and Anushree Roy. Precision measurement of the Casimir force from 0.1 to 0.9 μm. Physical Review Letters 81 (1998) 4549–4552. http://arxiv.org/pdf/quant-ph/0106045v1.pdf, queried on July 17, 2013

[9.4] H. B. Chan, V. A. Aksyuk, R. N. Kleiman, D. J. Bishop and Federico Capasso. Nonlinear micromechanical Casimir oscillator. Physical Review Letters 87 (2001) 211801-1 [4 pages]. http://www.nature.com/nphoton/journal/v5/n4/fig_tab/. nphoton.2011.39_F6.html, queried on July 17, 2013

[9.5] J. Knödlseder, P. Jean, V. Lonjou et al. The all-sky distribution of 511 keV electron-positron annihilation emission. Astronomy and Astrophysics 441 (2005) 513–532

[9.6] Yuri. I. Izotov and Trinh. X. Thuan. The primordial abundance of ^{4}He revisited. The Astrophysical Journal 500 (1998) 188–216

[9.7] NASA sun fact sheet. http://nssdc.gsfc.nasa.gov/planetary/factsheet/sunfact.html, queried on October 16, 2011

[9.8] P. Kervella, F. Thévenin, P. Morel, P. Bordé and E. Di Folco. The interferometric diameter and internal structure of Sirius A. Astronomy and Astrophysics 408 (2003) 681–688

[9.9] P. Eggenberger, C. Charbonnel, S. Talon, G. Meynet, A. Maeder, F. Carrier and G. Bourban. Analysis of α Centauri AB including seismic constraints. Astronomy and Astrophysics 417 (2004) 235–246

[9.10] P. Kervella, F. Thévenin, P. Morel, G. Berthomieu, P. Bordé and J. Provost. The diameter and evolutionary state of Procyon A. Astronomy and Astrophysics 413 (2004) 251–256

[9.11] J. L. Provencal, H. L. Shipman, Detlev Koester, F. Wesemael and P. Bergeron. Procyon B: Outside the iron box. The Astrophysical Journal 568 (2002) 324–334

[9.12] Saul J. Adelman, H. Caliskan, D. Kocer, H. Kablan, K. Yüce and S. Engin. Elemental abundance analyses with DAO spectrograms. Astronomy and Astrophysics 371 (2001) 1078–1083

[9.13] R. E. M. Griffin and A-E-Lynas-Gray. The effective temperature of Arcturus. The Astronomical Journal 117 (1999) 2998–3006

[9.14] C. A. Hummel, J. T. Armstrong, A. Quirrenbach et al. Very high precision orbit of Capella by long baseline interferometry. The Astronomical Journal 107 (1994) 1859–1867

[9.15] I. A. Usenko, A. S. Miroshnichenko, V. G. Klochkova and M. V. Yushkin. Polaris, the nearest Cepheid in the galaxy : Atmosphere parameters, reddening and chemical composition. Monthly Notices of the Royal Astronomical Society 362 (2005) 1219–1224

[9.16] T. P. Walker, G. Steigman, K. Ho-Shik, D. M. Schramm, K. A. Olive. Primordial nucleosynthesis redux. The Astrophysical Journal 376 (1991) 51–69

[9.17] Albrecht Unsold. Uber die Temperatur der Sonnenkorona. Zeitschrift fur Astrophysik 50 (1960) 48–56

[9.18] J. W. Brosius, J. M. Davila, R. J. Thomas and W. T. Thompson. Solar coronal temperature diagnostics using emission lines from multiple stages of ionization of iron. The Astrophysical Journal 425 (1994) 343–347

[9.19] R. Kariyappa, E. E. DeLuca, S. H. Saar, L. Golub, L. Damé, A. A. Pevtsov and B. A. Varghese. Temperature variability in x-ray bright points observed with Hinode/XRT. Astronomy and Astrophysics 526 (2011) A78 [6 pages]

[9.20] Grote Reber. Cosmic static at 144 meters wavelength. Journal of the Franklin Institute 285 (1968) 1–12

[9.21] S. Prawer and R. J. Nemanich. Raman spectroscopy of diamond and doped diamond. Philosophical Transactions of the Royal Society A 362 (2004) 2537–2565

[9.22] Xiaochen Dong, Y. Shi, Y. Zhao et al. Symmetry breaking of graphene monolayers by molecular decoration. Physical Review Letters 102 (2009) 135501-1 [4 pages]

[9.23] PingHeng Tan, YuanMing Deng and Qian Zhao. Temperature dependent Raman spectra and anomalous Raman phenomenon of highly oriented pyrolytic graphite. Physical Review B 58 (1998) 5435–5439

[9.24] M.J. Matthews, M.A. Pimenta, G. Dresselhaus, M.S. Dresselhaus and M. Endo. Origin of dispersive effects of the Raman D band in carbon materials. Physical Review B 59 (1999) 188–216

[9.25] I. Pocsik, M. Hundhausen, M. Koos and L. Ley. Origin of the D peak in the Raman spectrum of microcrystalline graphite. Journal of Non-Crystalline Solids 227–230 (1998) 1083–1086

[9.26] Marie Foret, Eric Courtens, René Vacher and Jens-Boie Suck. Scattering investigations of acoustic localization in fused silica. Physical Review Letters 77 (1996) 3831–3834

[9.27] R. C. Zeller and R. O. Pohl. Thermal conductivity and specific heat of noncrystalline solids. Physical Review B 4 (1971) 2029–2041

10 The radiation formula of Max Planck

> Without speculation there is no good and original observation.
> (*Charles Darwin*)

On the basis of classical laws, it is not possible to describe the spectral intensity distribution of heat radiation of a body of temperature T. Max Planck, first in a purely speculative way, therefore presented an "interpolation formula" for the thermal radiation of a body of temperature T, which agreed very well with measured data. From this he deduced that the energy of vibrating charges is a discrete variable and thus always an integer multiple of a smallest energy quantum. The radiation from a radiating body is not emitted uniformly and continuously, but in small energy portions, which are dependent on frequency. Hence, the temperature radiation consists of a large number of many independent, short wave trains with a specific energy.

Heat radiation is independent of the size and shape of the sample. Also, the nature of the substance does not matter. The Planckian radiation formula is therefore universal. It is a thermodynamic system of a countably infinite number of oscillators that exchange radiation with the environment.

The independence of the blackbody radiation on material properties has far-reaching consequences, because it imperatively demands that not only radiation oscillators, but all matter can absorb or release energy only in quantized form. If matter with energies other than hf is present, it is not observable at all due to the hypothesis. The Planckian radiation formula is a radical break with the classical idea that physics can only be described with differential equations.

According to Max Planck, the frequency-dependent, spectral energy density u(f) with the dimension $J\,m^{-3}\,Hz^{-1}$ can be parameterized in the frequency interval [f, f + df] by

Formula 10.1 $u(f) = 8\pi\, h\, f^3\, c^{-3}/(\exp\{hf/k_BT\}+\delta) = n(f)\, hf$

Formula 10.2 $n(f) = 8\pi\, f^2\, c^{-3}/(\exp\{hf/(k_BT)\}+\delta)$

The parameter δ determines the type of the energy distribution. With $\delta = -1$, a Bose–Einstein distribution, and with $\delta = +1$, a Fermi–Dirac distribution is described. With $\delta = 0$, the classical Maxwell–Boltzmann distribution is also included, which leads to the Rayleigh–Jeans law of classical electrodynamics, which provides useful results for high temperatures and large wavelengths. In the derivation, it was assumed that the radiation source radiates uniformly in all spatial directions over the full solid angle of 4π sr. The spectral energy density u(f) over the full solid angle has the dimension of an energy per unit volume and per unit frequency. It is synonymous with the number of oscillators (modes) with the energy hf per volume and per frequency in the frequency interval df. By Formula 10.1, the wave image, and by Formula 10.2, the photon image or particle image are embodied in a closed, cubic space.

https://doi.org/10.1515/9783110612387-010

The total photon density (total number of photons per volume) and the total energy density (total energy per volume) are obtained by integrating the frequency distribution of the photon density n(f) or the spectral energy density u(f) over all frequencies from zero to infinity.

For $\delta = \pm 1$, the particle number densities n (unit m^{-3}) are given by

Formula 10.3 $n_{Planck}(\delta = -1) = zeta(3)\, 2^4\, \pi\, k_B^3\, (hc)^{-3}\, T^3$
$n_{Planck}(\delta = -1) \approx 20.3 \cdot (T/K)^3\, cm^{-3}$

Formula 10.4 $n_{Planck}(\delta = +1) = 3/4 \cdot n(\delta = -1)$

zeta (s) is the Riemann zeta function, which for s = 3 amounts to ≈1.202 056.

For $\delta = \pm 1$, the energy densities $(\varepsilon/V)_{Planck}$ (unit $J\, m^{-3}$) are given by

Formula 10.5 $(\varepsilon/V)_{Planck}(\delta = -1) = 2^3\, \pi^5\, 15^{-1}\, k_B^4\, (hc)^{-3}\, T^4$
$(\varepsilon/V)_{Planck}(\delta = -1) \approx 7.57 \cdot 10^{-16}\, (T/K)^4\, J\, m^{-3}$

Formula 10.6 $(\varepsilon/V)_{Planck}(\delta = +1) = 7/8 \cdot (\varepsilon/V)_{Planck}(\delta = -1)$

If in Formula 10.3 the particle number density $n_{Planck}(\delta = -1)$ is equated with the particle number density n of Table 5.1, a temperature T_n is established. It is given by

Definition 10.1 $T_n \equiv k_B^{-1}\, h\, c\, n^{1/3}\, zeta(3)^{-1/3} \pi^{-1/3}\, 2^{-4/3} \approx 5.52 \cdot 10^5\, K$

which in units of $T(r = a_H)$ is equivalent to

Relation 10.1 $T_n/T(r = a_H) = 2^{12}\, \pi^{-55/9}\, zeta(3)^{-1/3} \approx 3.528 \approx 7/2$

if $\alpha = \alpha_{geom}$ and $m_e = m_{e\_geom}$ according to Relation 6.4 are used. Are T_n or $T(r = a_H)$ the relevant energy scales of neutron stars, or giant nuclei, respectively? The lonely, radio-quiet neutron star RX J1856.5-3754 in the constellation of Corona Australis is probably the closest neutron star known with a surface temperature of $(5.5 \pm 1.5) \cdot 10^5$ K depending on stellar models.

An energy kr^2 can be provided using the relationship $\omega^2 = k/m$ and the equivalence relationships of particle physics $f = mc^2/h_{bar}$ and $r = h_{bar} / (mc)$. Putting the three terms into kr^2 yields $mc^2(2\pi)^2$ after simple algebra. If $mc^2(2\pi)^2$ is then equated with $k_B T_n/2$ due to the uniform distribution theorem, a value of ≈0.60 eV/c^2 for m is obtained, which is close to $m_\nu \approx 0.62\, eV/c^2$ defined by Relation 6.17.

If the value $(8\pi)^{-1}$ from Relation 6.1 is used in Formula 10.5 for the energy density $(\varepsilon/V)_{Planck}(\delta = -1)$, a temperature T_{rho} is established, which can be calculated by

Relation 10.2 $(8\pi)^{-1} = 2^3\, \pi^5\, 15^{-1}\, k_B^4\, (hc)^{-3}\, T_{rho}^4$

The thus calculated temperature T_{rho} is defined by

Definition 10.2 $T_{rho} \equiv 2^{-3/2} \pi^{-3/2} 15^{1/4} k_B^{-1} (hc)^{3/4}$

and amounts to ≈2,693 K. Does T_{rho} possibly have meaning for stars as a critical boundary?

10.1 The diffuse microwave background radiation and the connection to the reference temperature

From all directions of the universe a remarkably uniform microwave radiation or cosmic microwave background (CMB) can be detected. The WMAP (Wilkinson Microwave Anisotropy Probe) spacecraft accurately determined a full-sky map of this microwave background radiation of the universe. This mission was mainly developed to measure temperature differences across the sky in the CMB. Already in 1992, the precursor satellite COBE (Cosmic Background Explorer) provided the first blurred images of this homogeneous background radiation filling the universe in every direction. The uniformity of the microwave background radiation and the chemically uniform structure are clear indications of isotropy in the universe. Such a homogeneity implies that all areas of space must be in a causal relationship. All measured data reflect a thermal radiation that corresponds to an almost perfect Planckian blackbody radiation which can be parameterized by a single temperature. According to the WMAP [10.1] measurement, this temperature is 2.725 K. According to a refined analysis of data from FIRAS on board of COBE [10.2], it is **(2.728 ± 0.004) K** with a 95% confidence level.

The uniformity and strength of the microwave background radiation from one edge of the observable universe to the other is difficult to understand because the two edges are several million light-years apart. How was thermal equilibrium reached? This "horizon problem" poses serious problems for cosmologists. Today, it is neatly explained by the theory of inflation, upon which not all physicists agree, since it solves certain problems, but additionally causes other discrepancies. Although it cannot be said that the inflationary model of the universe is confirmed, alternatives still attract little attention.

Is the Standard Model, proposed by Alan Guth, really the only scenario possible to explain why the universe is flat and homogeneous with a smooth distribution of matter and radiation on large scale? In the Standard Model, the temperature of the microwave background radiation is a parameter, which must experimentally be determined and cannot be calculated by a theory. It is indisputable that the microwave background radiation is one of the most important astronomical observations at all.

In addition to the postulate that light is quantized, Max Planck also used thermodynamic considerations to understand the blackbody radiation. Due to the fact that the microwave background radiation is an almost perfect blackbody radiation, thermodynamics as a doctrine of energy conversions has to play a role. Does the universe have more in common with thermodynamics than we suspect? A thermal equilibrium in itself requires only a few parameters to describe it, but which ones? With

Hypothesis 10.1 $T_{CMB} \equiv T_\gamma = T(r{=}\Lambda)\ 2^{1/2}$

using CODATA values for the natural constants, a value of ≈**2.7277 K** is obtained for the microwave background temperature T_{CMB}. Is it not amazing that the temperature T_{CMB}, calculated according to Hypothesis 10.1, corresponds with a fantastic accuracy to the microwave background temperature determined from the COBE data? The match is almost perfect, almost too perfect.

The factor $2^{1/2}$ was empirically determined, and it is interesting that already in Hypothesis 3.1 the same factor appeared. Hypothesis 10.1 somehow establishes a connection between thermodynamics and gravitation, since the length Λ and the mass m_0 are related by Definition 2.1. But, why does Hypothesis 10.1 allow such an exact description of the measured microwave background temperature of the observable universe? Is $w(r = \Lambda)$ the energy source of the microwave background radiation? What is the origin of the factor $2^{1/2}$?

Can the temperature $T(r = \Lambda)$ possibly be understood as the temperature T_ν of the neutrino background radiation that is being intensely searched today? Currently, there are no observational hints for the existence of such a background radiation. According to current doctrine the relationship $T_\nu = (4/11)^{1/3}\ T_\gamma$ is valid for massless neutrinos suggesting, because $(11/4)^{1/3}$ amounts to ≈1.401, that the temperature $T(r = \Lambda)$ could actually correspond to the neutrino temperature T_ν. Theoretically it is expected that in the relationship $T_\nu = (4/11)^{1/3}\ T_\gamma$ the proportionality factor for neutrinos with mass is a little bit smaller, what $2^{-1/2}$ compared to $(4/11)^{1/3}$ undoubtedly is.

The temperature $T(r = \Lambda)\ 2^{1/2}$ can be written as a function of the four natural constants c, h, m_e and k_B. It is given by

Relation 10.3 $T_{CMB} \approx T(r = \Lambda)\ 2^{1/2} = c^{8/9}\ h^{2/3}\ m_e^{1/9}\ k_B^{-1}\ \alpha^{4/3}\ 2^{-19/6}\ \pi^{-38/27}$

If the expression $c^{8/9}\ h^{2/3}\ m_e^{1/9}\ k_B^{-1}$ in Relation 10.3 is replaced by the reference temperature τ and the Sommerfeld constant α by $\alpha_{geom} \equiv 2^{-6}\ \pi^{-2/3}$, then

Relation 10.4 $T(r = \Lambda) = 2^{-5}\ c_1^{3}\ \tau \quad (=T_\nu\ ?)$

follows, from which

Relation 10.5 $T(r = \Lambda)\ 2^{1/2} = 2^{-9/2}\ c_1^{3}\ \tau \approx T_{CMB}$

for T_{CMB} can be derived.

10.2 The diffuse background radiation at high energies and the connection to the energy scale of nuclear binding

Is there a background temperature, which is responsible for a uniform electromagnetic radiation of highest frequencies and energies? In other words, which process can transform gravitational energy or nuclear energy into hard x-ray radiation?

X-ray instruments aboard of various spacecrafts, for example on board the HEAO-1 (High Energy Astronomical Observatory), have measured the extragalactic, diffuse x-ray background (CXB) of regions of the sky without point sources [10.3]. Without spatial structures, most of the x-ray energy is concentrated between 10 and 100 keV with a distinct maximum around ≈30 keV. The spectrum [10.4] has a characteristic bell shape and shows a progression, which steadily increases between 3 to about 20 keV, then flattens and continuously decreases again between 40 and 50 keV. An analytical fit to the measured spectrum of HEAO-1 by D. E. Gruber and coauthors [10.3] gave a maximum at ≈29.3 keV.

The origin of the entire cosmic, diffuse x-ray background is still unclear today. The astronomers' Standard Model interprets it either as a result of thermal bremsstrahlung of a hot, intergalactic gas, or as the cause of the superposition of x-ray radiation from individual sources that are so far away that they cannot be resolved by the telescopes available today and therefore merge into a uniform background. A thermal bremsstrahlung or x-ray radiation with a continuous emission spectrum similar to the blackbody radiation arises, when in an ionized gas plasma thermal electrons of high temperature are scattered and decelerated in the Coulomb field of positive ions (matter particles). If a Maxwell–Boltzmann distribution is assumed for the distribution of the electrons, the intensity in the spectrum drops exponentially.

The field energy $w\{r = L_2(\Lambda)\}$ is ≈46.1 keV. Could $w\{r = L_2(\Lambda)\}$ be the relevant energy scale for the diffuse x-ray background, as $w(r = \Lambda)$ is for the microwave background? What is the origin of the maximum in the x-ray background at ≈30 keV? In Section 8.2 it was found that $h_{bar}\, c\, L_2(\Lambda)^{-1}$ represents a relevant energy scale of the nuclei, and the average nuclear binding energy amounts to about $h_{bar}\, c\, L_2(\Lambda)^{-1}\cdot(2/\pi)$. Interestingly, multiplying the field energy $w\{r = L_2(\Lambda)\}$ by the same factor $2/\pi$ yields ≈29.4 keV. Is it a numerical coincidence that the maximum of the diffuse x-ray background peaks at $w\{r = L_2(\Lambda)\}\cdot(2/\pi)$?

If the field energy $w\{r = L_2(\Lambda)\}$, under the constraint that the total momentum is conserved, is totally used to produce two photons of wavelength λ_{CXB}, energy conservation requires that

Definition 10.3 $w\{r = L_2(\Lambda)\} = 2\, h\, c\, \lambda_{CXB}^{-1}$

applies. With CODATA values for the natural constants, this results in

Relation 10.6 $\lambda_{CXB} \approx 0.538\ \text{Å} \approx 1.016\ a_H$

Remarkably, λ_{CXB} is approximately the Bohr radius a_H. The ratio λ_{CXB}/a_H is dimensionless, and taking into account α_{geom} and m_{e_geom} must yield an expression dependent only on powers of the number constants 2 and π. By algebraic transforming

Relation 10.7 $\lambda_{CXB}/a_H = 2^{-11}\,\pi^{20/3} \approx 1.0069$

results. A comparison of Relation 10.7 with Relation 9.14 provides

Relation 10.8 $\lambda_{CXB}/a_H \equiv r_e/(4\pi\, r_{grav})$

which relates four characteristic lengths to each other.

10.3 The interdependence of h, c and the Boltzmann constant k_B

In solving the problem of blackbody radiation and adapting it to experimental facts, Max Planck has phenomenologically introduced the two constants h and k. The constant k named Max Planck in honor of Ludwig Boltzmann the Boltzmann constant k_B. The thermodynamic temperature T and Boltzmann's constant k_B are found in all physical laws only as product k_BT with the unit of energy, and in classical thermodynamics everything can be attributed to such a product. Accordingly, only the product k_BT and never the constant k_B itself is measured. As a consequence, the Boltzmann constant itself never appears in a dimensionless quantity. This property clearly differentiates it from the other constants of nature. It is believed [10.5] that it combines microscopic, individual events with collective behavior because of the relationship $R_{gas} = k_B\, N_{Avogadro}$. Or does it play an analogous role as the proportionality constant h between energy and frequency, or the proportionality constant c^2 between energy and mass? There is some controversy in the scientific community because the physical origin of k_B is still not clear. What is certain is that the three quantities k_B, h and c are the constants on which the Planckian radiation formula and the relativistic quantum field theory are based.

According to its units, the Boltzmann constant k_B is nothing more than a proportionality constant, which converts thermal energy with the unit J into a temperature with the unit K, or vice versa. By such a definition, it is closely related to the temperature measurement and the idea that it relates the average kinetic particle energy and the thermal energy of a system without macroscopic motion. As long as the temperature is not considered a fundamental physical quantity, the Boltzmann constant remains a pure conversion constant and could take any values without affecting physics. Is this widespread opinion right?

Due to metrological and principal considerations, 1 K is formally defined as the exact fraction 1/273.16 of the thermodynamic temperature of the triple point of pure water. By locating the second temperature fix point at the triple point of water, it is thus associated with a rather accidental material property. Although water is the fluid that plays the most important role on earth's surface, it is meaningless in the universe. The absolute calibration is carried out using primary thermometers or gas thermometers whose temperature dependence can be described by fundamental relationships of physics. The experimental calibration is tedious and difficult, the experiments are very complicated and it raises the question how accurate the temperature scale is defined by this method.

Another method is the measurement of the gas constant R_{gas} using acoustic gas thermometry and the calculation of the Boltzmann constant k_B by means of the relationship $k_B = R_{gas}/N_{Avogadro}$. The latter is also the method used today to determine the tabulated CODATA value. This approach is based on the phenomenological equation of state of the ideal gas. However, it is by no means obvious that the gas constant R_{gas} based on the theory of statistical mechanics has the same value for all ideal gases. The postulate of the independence of the gas constant R_{gas} was first established by Amadeo Avogadro. The constant k_B is definitely an independent measure today that has no reference to other fundamental constants.

The question arises as to whether the Boltzmann constant can be freely chosen without any impact on physics. Or is there possibly a connection with other natural constants? From a metrological point of view, the constants h and k_B must somehow be related, since extreme temperatures can only be determined by measuring the frequency and the spectral intensity of emitted photons. It is also a fact that both h and k_B are measures of stochastic system properties. In any case, according to the current opinion, the status of k_B with respect to h and c is not fully clear.

If in Relation 5.3 the Sommerfeld constant α and the mass m_e are replaced by α_{geom} and m_{e_geom}, the simple

Relation 10.9 $$k_B\tau = 2^{-17/4}\,(ch)^{3/4}$$

remains. It is astonishing that by means of the assumptions made for α and m_e, the irrational number π cancels in the expression for the calculation of the characteristic heat energy $E_\tau = k_B\tau$ and loses any meaning for the physical world. Is this numerology, or is it recognizing that the assumptions made are correct, since otherwise the number constant π would not have been canceled in the calculation of $k_B\tau$? Is it a coincidence that the number constant π is also canceled in the case of the nine-dimensional hypersphere $V_9^{-1}(c_1)$? For the unit of $c^{3/4}\,h^{3/4}$

Relation 10.10 $$[c^{3/4}\,h^{3/4}] = m^3 = J$$

holds, if the relation $kg = m\,s^2$ is used.

Should for τ a natural number arise similar to the nine-dimensional hypersphere $V_9^{-1}(c_1)$? In other words, is the Boltzmann constant such that the reference temperature τ should correspond exactly to a natural number with the unit Kelvin? A bold assumption with the possibility of thereby defining the Boltzmann constant axiomatically? For those who regard the Boltzmann constant as a man-made, accidentally chosen conversion factor between thermal and mechanical energies, such a definition would certainly not be a problem. The path outlined would thereby undoubtedly link the Kelvin with other known, fundamental constants of nature without distinguishing a temperature fix point or a measurement method, and would be free from any human subjectivity. Its simplicity and the fact that the definition is independent of space, time and material speak for the expediency of the determination thus made. Likewise, the newly defined temperature values would not be very different from the old ones.

The question naturally arises, which natural number should be chosen for τ. A comparison with the number 1,890 found in Section 3.2 suggests that because of

Relation 10.11 $1{,}134 = 3/5 \cdot 1{,}890 = 3/5 \cdot V_9^{-1}(c_1)$

the number 1,134 could make sense. The mean kinetic energy per particle $<\varepsilon>$ in a three-dimensional, nonrelativistic quantum system with energy ε and momentum p related by $\varepsilon = cp^2$ is given by

Formula 10.7 $<\varepsilon> = (3/5) \cdot \varepsilon_{Fermi}$

which reveals a remarkable formal similarity to Relation 10.11 and ensures that the calculation of the Boltzmann constant k_B as a measure of randomly distributed energy is based on a statistical method.

As a result of the foregoing considerations, the Boltzmann constant would not simply represent a proportionality constant or scaling parameter for another energy scale, but would actually be an autonomous quantity with a statistical background. If so, the thermal equilibrium between different degrees of freedom interacting with each other plays a fundamental role. In whatever way these interactions of different ranges take place, the heat energy and thus the otherwise isolated temperature together with the Boltzmann constant obtains a universal status. Equally distributed energy also implies that the energy density must be the same everywhere, and therefore any charge separation in space is neutralized somehow.

Assigning τ exactly the value 1,134 K,

Hypothesis 10.2 $k_{B\_geom} \equiv (ch)^{3/4}\, 2^{-17/4}\, 1{,}134^{-1} \cdot K^{-1} = (ch)^{3/4}\, 2^{-17/4}\, 5/3\, V_9(c_1) \cdot K^{-1}$

can be derived by means of Relation 10.9. It allows to calculate the Boltzmann constant k_B on the basis of the constants h and c alone. The geometrized Boltzmann constant k_{B_geom} amounts to $\approx 1.379 \cdot 10^{-23}$ J K^{-1}, if for the Planck constant h and the speed of light in vacuum CODATA values are used. The deviation from the CODATA value is about 0.1%.

In the current SI system, only the "core units" kg, s and Kelvin are independent. All other units, such as the meter or the ampere, are units derived from the units of kg and s. The Kelvin is the only unit with no link to other units. The question therefore arises whether it would not be better to fix the Boltzmann constant k_B as an atomic entropy unit – such as the vacuum speed of light or the magnetic field constant – to a value without measurement error. This would finally give statistical thermodynamics the meaning it deserves. This implied that not h, but k_B is the origin of the energy uncertainty, that is, k_B is responsible for the fact that small impacts can no longer be distinguished from chaos and a deterministic description of microscopic phenomena is not possible. In this view, c and k_B are conversion factors that can be arbitrarily set. At least physically, this makes sense, since both c and k_B are based on completely different physical concepts. Regardless of whether h or k_B is fixed, Hypothesis 10.2 suggests that both quantities are interrelated, and probably none should be fixed without the other for consistency reasons.

References

[10.1] Five year result on the oldest light in the universe. http://map.gsfc.nasa.gov/news/5yr_release.html, queried on October 16, 2011

[10.2] D.J. Fixsen, E.S. Cheng, J.M. Gales, J.C. Mather, R.A. Shafer and E.L. Wright. The cosmic microwave background spectrum from the full COBE FIRAS data set. The Astrophysical Journal 473 (1996) 576–587

[10.3] D. E. Gruber, J. L. Matteson and L. E. Peterson. The spectrum of diffuse cosmic hard x-rays measured with HEAO 1. The Astrophysical Journal 520 (1999) 124–129. http://cds.cern.ch/record/383660/files/9903492.pdf, queried on June 30, 2013

[10.4] R. Gilli. The x-ray background and the deep x-ray surveys. http://arxiv.org/pdf/astro-ph/0303115v1.pdf, queried on June 30, 2013

[10.5] Robert Rompe. Was sind und was bedeuten die Elementarkonstanten? Annalen der Physik 42 (1985) 559–576

11 The gravitational fine-structure constant α_{grav} as a number constant and the connection to α

To understand the "essence of gravity" in physics means to relate gravitation to other interactions. (*H. J. Treder*)[1]

The electromagnetic interaction with the coupling constant $\alpha^{1/2}$ and gravity obey the same inverse-square distance law. This coincidence has strongly influenced physical thinking since the discovery of the Coulomb law. If gravitation is interpreted as an interaction analogous to electromagnetism, it is natural to construct a dimensionless coupling constant $\alpha_{grav}{}^{1/2}$ of gravity using natural constants in accordance with Formula 5.4. However, it is necessary to define a mass scale (elementary mass) so that a pure number is obtained for $\alpha_{grav}{}^{1/2}$. To clarify this deep physical question,

Definition 11.1 $\alpha_{grav}{}^{1/2} \equiv m_{grav}/m_{Planck}$ with $m_{Planck}{}^2 = h_{bar}\, c\, G^{-1}$

shall apply. The Planck mass m_{Planck} is ≈21.8 µg. Its calculation involves only h, c and G and not the mass of a particle. It is the only unit of the Planck units that is not as unimaginably small as all the other Planck units. By Definition 11.1, m_{grav} is designated as an elementary mass or an invariant unity of the material mass. It thus plays an analogous role as the elementary charge q_e in the electromagnetic interaction and has atomistic properties, which means that it must be a conserved quantity of matter.

Due to Definition 11.1, all gravitational forces, like the Coulomb forces, can be considered as integral multiples of a discrete, quantized mass. If for the mass m_{grav} the value of $\approx 1.574 \cdot 10^{-27}$ kg resulting from Relation 6.21 is used, 1 kg contains $\approx 6.4 \cdot 10^{26}$ elementary masses. Both the elementary charge q_e and the elementary mass m_{grav} can only be detected by forces acting between them. In contrast to the elementary charge, no elementary mass m_{grav} has been observed as a source of gravity until today. For this reason, the coupling constant $\alpha_{grav}{}^{1/2}$ of gravitation is rarely mentioned in the physical literature because it depends on the accidental choice of the "gravitational charge" m_{grav}. Often the mass of the stable electron or the mass of the stable proton is chosen, for which, apart from their stability, there is no justifiable, rational reason. In any case, there is no agreement in the literature on which value should be chosen for m_{grav} in Definition 11.1.

By transforming Definition 11.1,

Relation 11.1 $\alpha_{grav} = (m_{grav}/m_{Planck})^2 = h_{bar}{}^{-1}\, c^{-1}\, G\, m_{grav}{}^2$

1 Astronomische Nachrichten 304 (1983), page 149.

https://doi.org/10.1515/9783110612387-011

can easily be deduced. The number α_{grav} can also be considered as the ratio of the interaction energy or binding energy of two particles of mass m_{grav} at a distance r_{grav} to the rest energy $m_{grav}\,c^2$. This is valid because

Relation 11.2 $(G\, m_{grav}^3\, c\, h_{bar}^{-1}) / (m_{grav} c^2) = h_{bar}^{-1}\, c^{-1}\, G\, m_{grav}^2 = \alpha_{grav}$

and

Relation 11.3 $G\, m_{grav}^2 / r_{grav} = \alpha_{grav}\, (m_{grav}\, c^2)$

apply, which can be derived using Relation 6.20. Both Relations 11.2 and 11.3 are but different views of the same facts. By Definition 11.1, the gravitational fine-structure constant depends on the fundamental natural constants h, c, G, and the elementary mass m_{grav}. For α_{grav}, in contrast to the Sommerfeld constant α, no pure number results solely from the combination of natural constants. In addition, the mass m_{grav} axiomatically fixed by Equation 6.2 must be selected so that a number without dimensions is obtained. This allows then to attribute the gravitational fine-structure constant to the fundamental constants h, c, m_e and G. Using m_{grav} from Equation 6.2, a numerical value for α_{grav} of $\approx 5.2 \cdot 10^{-39}$ results.

11.1 Formulaic connection between astronomy and atomism

The reciprocal gravitational fine-structure constant α_{grav}^{-1} can also be written as

Relation 11.4 $q_e^2 / (4\pi\varepsilon_0\, G\, m_{grav}^2) = \alpha_{grav}^{-1}$

This establishes a link to the Eddington–Dirac number, which instead of m_{grav}^2 contains the product $m_e \cdot m_p$. The Eddington–Dirac number represents the ratio of the Coulomb force to the gravitational force of electron and proton. This ratio is independent of the distance between electron and proton, since both particles obey the same force law as a function of distance. For the first time Eddington and Dirac pointed out that "large numbers" must determine the scale of astronomical parameters.

If geometrized values are used for m_e and for α, simple algebraic transformations of Relation 11.1 produce

Relation 11.5 $\alpha_{grav}^3 = h\, c^{-7}\, G^2\, 5\, 3^{-1}\, 2^{-4}\, \pi^{-2}$

It should be noted that for the unit kg, the synonymous unit m s^2 must be used so that the units on the right side of Relation 11.5 cancel and a dimensionally consistent relationship is obtained. The term $h^{-1}\, c^7\, G^{-2}$ is also referred to as the Planck energy density.

It becomes interesting when Relation 11.5 is linked to Relation 6.23, which provides a geometrized value also for G. That is to say, it allows to eliminate the product $h\, c^{-7}\, G^2$ in Relation 11.5. Some algebraic transformations then give the "aesthetic"

Relation 11.6 $3/5 \cdot \alpha_{grav} = 2^{-9} \pi^{-72} = \{2 \pi^8\}^{-9} = \{2^{71} \alpha_{geom}{}^{12}\}^9$

which is rid of all natural constants and numerically unites the gravitational fine-structure constant and electromagnetism, and thus establishes a formal relationship between astronomy and atomism first promulgated by the Greeks.

Interestingly, the dimension nine found in Section 3.2 crops up again. A remarkable coincidence? Or does the nine-dimensional hypersphere indeed play a role in the symbolic description of our world? Relation 11.6 is of very simple form and undoubtedly satisfies the principle of simplicity of hypothesis. It confirms in some way the choice of the "gravitational charge" m_{grav} in the definition of the gravitational fine-structure constant.

Without the reduction (geometrization) of the electron mass to the constants h and c, such a simple expression for the "large number of Dirac" would certainly not have arisen, and the "arbitrary" definition of the electron mass by Relation 6.4 appears thereby in a completely different light. But simple mathematics does not necessarily mean that it is right. Does Relation 11.6 reflect the fact that gravitation, since α_{geom} is an electromagnetic constant, is generated by electromagnetism and must therefore be considered from this point of view? If this is true, it can formally be concluded from Relation 11.6 that gravitation, as experimentally observed, must be attributed to electromagnetism with only one algebraic sign.

By using the definitions of r_{grav}, L_{Planck} and L_{large}, it can easily be verified that the three length scales are linked by

Relation 11.7 $r_{grav}{}^4/(L_{Planck}\, L_{large})^2 = (3/5)\, \alpha_{grav}$

to the gravitational fine-structure constant. Likewise,

Relation 11.8 $G\, m_{grav}/c^2 = \alpha_{grav}\, r_{grav}$

results by simple algebraic transformations. Replacing in Relation 11.8 m_{grav} by $h_{bar}/(r_{grav}c)$ and using Definition 6.4,

Relation 11.9 $r_{grav} = \alpha_{grav}{}^{-1/2}\, L_{Planck}$

follows. Substituting L_{Planck} from Relation 11.9 into Relation 11.7 then yields

Relation 11.10 $\alpha_{grav} = r_{grav}/\{(3/5)^{1/2} L_{large}\}$

which can be regarded as an analogy to the identity $\alpha = \lambda_{e\_bar}/a_H$. The quantity λ_{e_bar} corresponds to the reduced Compton length of the electron and a_H is the Bohr radius. The length $(3/5)^{1/2} L_{large}$ is approximately 1.6 times larger than the length $R_{horizon}$.

11.1.1 The idea of Ernst Mach

Dennis Sciama took up the critique of Ernst Mach on Newton's theory and suggested that the particle horizon $R_{horizon}$ and the total mass $M_{horizon}$ contained in it should be reflected in the Newton constant G. He suggested

Ansatz 11.1 $c^2/G = M_{horizon}/R_{horizon}$

which determines the total mass $M_{horizon}$, if the particle horizon $R_{horizon}$ is fixed, for example by Relation 6.8. But, there is no compelling physical argument requiring that the idea of Ernst Mach be true.

Using Relation 11.8 and Ansatz 11.1 yields

Relation 11.11 $c^2/G = M_{horizon}/R_{horizon} = \alpha_{grav}^{-1} (m_{grav}/r_{grav})$

which mathematically reflects the idea of Ernst Mach, Sciama, Zeldovich and others that local physical processes, dictated by the state of a spherical, finite volume, are the source of the universal gravitational field.

11.1.2 Neutron stars

From observations it is known that no astronomical objects with higher densities than the atomic nuclei exist. It is believed that this context between nuclear physics and gravitation is fundamental and could be reflected in neutron stars. As a comparable measure, the diameter of neutron stars could be useful. However, accurately determining the diameter of such objects is very difficult and is controversially discussed in the literature because precise measurements are lacking. To extrapolate nuclear parameters to the macrocosm, only fundamental parameters of matter such as m_{grav}, r_{grav} and α_{grav} can be examined. Could the Compton length $2\pi\, r_{grav} = h/(m_{grav}c)$ and the scale factor $\alpha_{grav}^{-1/2}$ play a role? Interestingly, the product $2\pi\, r_{grav}\, \alpha_{grav}^{-1/2}$ gives ≈19.4 km and approximates the diameters of neutron stars listed in the literature quite well. Do all neutron stars possibly have the same radius of $\pi\, r_{grav}\, \alpha_{grav}^{-1/2}$? What is the significance of the mass $m_{grav}\, \alpha_{grav}^{-3/2}$ of the order of $\approx 2.1 \cdot M_{sun}$?

11.1.3 The Dirac conjecture about the connection of α_{grav} with α_{geom}

The two dimensionless numbers $\alpha_{grav} = 5 \cdot 3^{-1}\, 2^{-9}\, \pi^{-72}$ and $\alpha_{geom} = 2^{-6}\, \pi^{-2/3}$ somehow connect relativity, quantum mechanics, gravitation and the properties of subatomic particles. Using

Relation 11.12 $\ln(\alpha_{grav}{}^{-1})/\alpha_{geom}{}^{-1} \approx 0.6421...$

as a measure, which numerically connects the physics of the small and the physics of the large, a number close to 1 is obtained. Relation 11.12 reflects a conjecture of P. A. M. Dirac that dimensionless numbers, which can be constructed from the important natural constants of the universe and atomic theory, should be connected by simple mathematical relations involving coefficients of the order of magnitude unity [11.1].

It is noticeable that the number 0.6421... is quite close to the number phi with the approximate value of 0.618. The difference, however, is too large to be seriously associated with it. If there is any connection between the heuristically chosen ratio and mathematics, a number with a closer agreement must be found.

Most real problems in physics can only be solved with nonlinear equations. Many of them are well described by Hamiltonians with few degrees of freedom. The two-dimensional, area-preserving Chirikov standard map for two canonical dynamical variables plays a universal role in the study of such issues. The discrete Hamilton system simulates the dynamics of a one-dimensional oscillator with an external drive of intensity (A). The coupled difference equations for the momentum (p) and position (q) are given by

Formula 11.1

$$q_{n+1} = q_n + p_{n+1}$$
$$p_{n+1} = p_n + A\cdot\sin(2\pi\cdot q_n) \quad (\text{mod } 1)$$

The quantity A is a statistical parameter that, if suitably chosen, provides periodic or chaotic solutions. For $A < 1$ periodic solutions dominate. As nonlinearity gets stronger, that is, as A increases, the solutions become more and more chaotic. K. Hirose and coworkers [11.2] observed numerically that in the region of $0.64037 < A < 0.65130$ complex structural conditions exist and two stable solutions occur. Is it a coincidence that the ratio $\ln(\alpha_{grav}{}^{-1})/\alpha_{geom}{}^{-1}$ lies in this interval and could be connected to the dynamics of discrete Hamiltonian systems?

11.2 The universal Fermi constant as a number constant and the connection to gravity

In 1934, Enrico Fermi set up a first quantitative theory of radioactive beta decay analogous to the electromagnetic interaction. He attributed the decay to a point interaction of four fermions with a dimensional coupling constant G_F of $\approx 1.436\cdot 10^{-62}$ J m³. Due to its dimension, the universal Fermi constant G_F is proportional to an energy times a volume. Today it is usually tabulated in the alternative form $G_F\cdot q_e{}^2/(h_{bar}c)^3$ with the unit eV^{-2}. Experimentally, it is mainly determined from the precise lifetime and the mass of the charged muon.[2] The fact that only one neutrino helicity participates in the contact interaction reduces the cross section to half. Instead of absorbing the factor $2^{-1/2}$ in the definition of the Fermi constant, this fact is historically taken into account by stating the factor explicitly. It is important to know that the value of the Fermi constant depends strongly on theoretical assumptions, that is, the Standard Model of particle physics with quantum electrodynamics (QED) corrections.

Shortly after the discovery of the beta decay, Wolfgang Pauli [11.3] proposed to connect the "weird" Fermi constant G_F (unit J m^3) to the Newton constant G (unit m^3 s^{-2} kg^{-1}). Thoughts about it can be found, for example, in reference [11.4]. For such a comparison to be meaningful, however, dimensionless parameters are required for both gravity and beta decay. While the gravitational fine-structure constant α_{grav} (Relation 11.6) is suitable for gravity, a physically meaningful approach must first be found for the beta decay. But how can a dimensionless Fermi constant be formed?

Similar to the Sommerfeld fine-structure constant $\alpha = q_e^2 (4\pi\varepsilon_0)^{-1} (h_{bar}c)^{-1}$, a dimensionless Fermi constant α_{Fermi} can be constructed for the universal beta decay, if an interaction radius r_F is additionally introduced to which the interaction can be related. If the Fermi constant G_F is divided by hc and r_F squared, the "strange" dimension of the Fermi constant can be avoided. This gives

Definition 11.2 $\alpha_{Fermi} \equiv (G_F/2^{1/2})/(hc)/r_F^{\,2}$

which converts the dimensional Fermi constant in a dimensionless number if the interaction radius r_F is known.

Let the dimensionless Fermi constant α_F axiomatically be defined as the fourth root of the gravitational fine-structure constant α_{grav}. Even if the Standard Model completely ignores gravitation, the two interactions α_{grav} and α_F shall be related by

Hypothesis 11.1 $\alpha_F \equiv \alpha_{grav}^{\;1/4}$

which mathematically correlates the long-range gravitation with the weak (radioactive) interaction, and allows to calculate by Definition 11.2 the interaction radius r_F according to

Relation 11.13 $r_F = \alpha_{grav}^{\;-1/8}\,(G_F/2^{1/2})^{1/2}\,(hc)^{-1/2}$

2 The mass of the muon amounts to ≈105.6 MeV/c^2, or ≈$1.883 \cdot 10^{-28}$ kg. Interestingly, the corresponding reduced Compton wavelength of ≈1.87 fm is of the order of the length $L_2(L)$, which is ≈1.82 fm.

if $\alpha_{Fermi} = \alpha_F$ is set. Using CODATA values for h, c and G_F, a value of ≈13.8 fm is obtained.

Does Hypothesis 11.1 reflect the Pauli conjecture, which connects microcosm and macrocosm and allows to infer from a cosmic quantity on a subatomic parameter, and to understand quanta by another view? It is noteworthy that the interaction radius r_F is on the order of magnitude of the scattering length $L_2(\Lambda) = (n\ \Lambda^2)^{-1}$ (≈15.6 fm), whose energy equivalent $h_{bar}c\ L_2(\Lambda)^{-1}$ plays a central role as a characteristic parameter for nuclear binding energies in Section 8.1. It is certainly physically meaningful that the interaction radius r_F is on the order of magnitude, whose energy scale determines the mean energy released when protons and neutrons are separated from the nuclei.

Hypothesis 11.1 yields a numerical value for the dimensionless Fermi constant α_F of $\approx 2.7\cdot 10^{-10}$, which is independent of natural constants including G_F. This universal, dimensionless number reflects the origin of the weak radioactive decay or the creation of baryonic matter. It seems that both fission and aggregation of matter are processes that, as suggested by Wolfgang Pauli, are based on the "misunderstood" gravity.

11.3 Particle lifetime ratios

The strength of the weak interaction can be estimated by comparing the lifetime of the radioactive decay of the charged pion ($\Pi^{\pm}$) with the electromagnetic decay of the neutral pion (Π^0) [11.5]. This interpretation is possible because the lifetimes of unstable particles depend on which dominant interaction channel they disintegrate. Charged pions disintegrate almost to 100% due to the weak or radioactive interaction into a muon and a muon neutrino. By contrast, almost 100% of neutral pions disintegrate due to the electromagnetic interaction into two photons. Both decays are well-documented experimental facts.

For particles at rest, the mean lifetime of charged pions ($\tau_{pion\_charged}$) is $(2.6033 \pm 0.0005)\cdot 10^{-8}$ s, according to the Data Particle Group (queried on March 31, 2012). For neutral pions at rest, the particle physicists state a mean lifetime $\tau_{pion\_neutral}$ of $(8.52 \pm 0.18)\cdot 10^{-17}$ s. By comparing the two decays, it can be concluded that the weak interaction must be weaker by a factor $\tau_{pion\_neutral}/\tau_{pion\_charged}$ (**≈3.2·10⁻⁹**) than the electromagnetic interaction. The term "weak interaction" historically has its origin in this small number.

The experimental lifetime ratio shall be explained by

Ansatz 11.2 $\tau_{pion\_neutral}/\tau_{pion\_charged} \approx \alpha_F/\alpha_{geom}^{1/2} = \alpha_{grav}^{1/4}/\alpha_{geom}^{1/2} \approx \mathbf{3.2\cdot 10^{-9}}$

Comparing the value obtained by Ansatz 11.2 with the experimental value of $\approx 3.2\cdot 10^{-9}$ suggests that the ratio $\tau_{pion\_neutral}/\tau_{pion\_charged}$ can be very well approximated by $\alpha_F/\alpha_{geom}^{1/2}$. Does this justify the assumption, made in Hypothesis 11.1, how the radioactive interaction α_F is connected to α_{grav}? Or is it just a coincidence, and α_F is only in this case a "convenient measure" for describing the radioactive interaction? The quantity $\alpha_{grav}{}^{1/4}/\alpha_{geom}{}^{1/2}$ is a number constant, that is, a ratio that is independent of any measurement system.

The neutron is the longest living, unstable particle of particle physics. According to the Data Particle Group (queried on March 31, 2012), the mean lifetime of neutrons is (881.5 ± 1.5) s outside the nucleus. This is a fairly long lifetime that cannot be understood with the weak interaction alone. The lifetime ratio $\tau_{pion\_neutral}/\tau_{neutron}$, which connects a timescale still imaginable to humans with a microscopic timescale of quantum theory, might help understand the interaction involved similar to the pion. For the ratio $\tau_{pion\_neutral}/\tau_{neutron}$, an experimental value of $\approx$**9.5·10**$^{-20}$ results, which could be associated with the square root of the gravitational fine-structure constant.

In fact, it is easy to verify that the number constant $(7/4\ \alpha_{grav})^{1/2}$, which in decimal notation amounts to $\approx$**9.6·10**$^{-20}$, closely approximates the lifetime ratio $\tau_{pion\_neutral}/\tau_{neutron}$. The "magical" ratio 7/4, which, for example, provides also a correct result when the primordial abundance of ^{4}He is calculated, seems to play an important role in this case, too, and it is amazing how well the experimental data can be reconciled with it. The fraction 7/4 is also a characteristic number in the parameterization of nuclear binding energies.

The interaction causing the decay of the neutron is apparently smaller by a factor of $\approx 9.5\cdot10^{-20}$ than the electromagnetic one. Since the number $(7/4\ \alpha_{grav})^{1/2}$ represents a good approximation for this factor, this implies, in analogy to $\tau_{pion\_neutral}/\tau_{pion\_charged}$, a value of $\approx(7/4\ \alpha_{grav}\alpha_{geom})^{1/2}/\alpha_{geom}{}^{1/2}$ for the ratio $\tau_{pion\_neutral}/\tau_{neutron}$. Does the number constant $(7/4\ \alpha_{grav}\alpha_{geom})^{1/2}$ provide an indication of why an unbound neutron decays on average after approximately 15 min?

11.4 Representation of the anomalous magnetic moment of the muon by means of coupling constants

As explained earlier, the kinematics of the muon decay is crucial in determining the Fermi constant G_F. Since G_F and α_F should describe the same physical principle, it is reasonable to suppose that the dimensionless Fermi constant α_F could also be important in the calculation of the anomalous magnetic momentum of the muon a_μ. In addition, using α_F in the calculation of $\tau_{pion\_neutral}/\tau_{pion\_charged}$, an excellent agreement with the experiment can be achieved. These are good prerequisites for trying

Ansatz 11.3 $a_{\mu\_theo} = \alpha_{geom}/(2\pi)+\{\alpha_F/(2\pi)\}^{1/2} \approx 0.001165870$

With this approach, which contains only two terms, a theoretical value is obtained which deviates only slightly from the CODATA value (queried on March 31, 2012) of 0.00116592069(60). Both the Sommerfeld constant α_{geom} and the dimensionless Fermi constant α_F, which can be determined by Relation 11.6 and Hypothesis 11.1, are pure number constants with physical reference to the electromagnetic or radioactive interaction. In addition to the dimensionless constants α_{geom} and α_F, which are both based on a_μ independent physical concepts, only the factor 2π occurs in Ansatz 11.3.

Undoubtedly, Ansatz 11.3 provides a much simpler approximation to a_μ compared to the complicated approach of the common doctrine, which calculates with infinities. What is the cause of the deviation from the measured value of 0.00116592069(60)? A missing correction δa_μ? As a correction, the value of the simple term $4 \cdot 3^{-1} \cdot \alpha_F \cdot \alpha_{geom}^{-1}$ would be of the correct order of magnitude, since it yields a value of ≈0.001165920 for a_{μ_theo}. More about this subject can be found in Section 15.5.5.

11.5 The Z and W bosons

S. L. Glashow, A. Salam and S. Weinberg unified the weak and electromagnetic interactions on the basis of abstract group-theoretical considerations. The former causes the so-called radioactive beta decay. The conversion of the neutron into a proton, an electron and an antineutrino is an example of this. Analogous processes with photons are much faster. How can there be a similarity? Nevertheless, S. L. Glashow, A. Salam and S. Weinberg found a way how the electromagnetic and the weak interactions mix by postulating virtual W and Z particles, over which the processes theoretically occur in the form of so-called loops in Feynman diagrams. But, they were not able to say which mechanism created the huge masses of the W and Z particles.

The masses of these W and Z particles are of the order of magnitude of a nucleus of middle size and, in the Standard Model, free parameters that can only be measured. In today's accelerators, these particles can even be created in collisions by pair production if sufficient energy is available. The mechanism that generates the huge mass of the intermediate W or Z particle has not yet been fully elucidated, since theory provides no explanation why the particle masses are as large as experimentally observed. In the unit kilogram, the mass m_Z of the extremely short-lived Z particle is $\approx 1.6256 \cdot 10^{-25}$ kg. It is comparable to the mass of the radioactive, long-lived isotope ^{98}Tc with 43 protons and 55 neutrons ($\approx 1.6258 \cdot 10^{-25}$ kg). Technetium (Tc) is the first, that is, lightest element in the periodic table, of which there are no stable isotopes.

In the quantum field theory of S. L. Glashow, A. Salam and S. Weinberg, the W/Z mass ratio is given by [Particle Data Group 2017]

Formula 11.2 $m_W/m_Z \equiv \cos(\theta_W) = \mathbf{0.88153} \pm 0.00017$

where θ_W denotes the electroweak mixing or Weinberg angle. The mass m_W is the mass (≈80.385 GeV/c^2) of the either positively or negatively charged $W^{+/-}$ boson, which is similar to the electron-positron pair, and the mass m_Z is the mass (≈91.1876 GeV/c^2)

of the neutral Z boson, which can be attributed to the photon. These two quantities have been determined by great experimental effort and, as averaged over many experiments, are quite accurate.

The mixing angle θ_W is a measure of the tendency of the particles to transform into one another. Often literature cites the quadratic sine value $\sin^2(\theta_W)$ of θ_W. The Standard Model of particle physics is unable to derive theoretically the W/Z mass ratio or the masses m_W or m_Z. Interestingly, in the physical literature, a number of the same order of magnitude is also not found in other contexts. What makes this fundamental ratio so unique?

Most interesting is the proximity of the W/Z mass ratio to the ratio $IE_{atomic}/IE_{molecular}$ of ≈**0.88163**, which was calculated in Section 6.2 by means of Ansatz 6.1 using geometrized quantities and experimental data, that is, the hyperfine splitting of the hydrogen ground state $(\Delta f)_H$. A still higher precision for the W/Z ratio is needed to better understand this astonishing coincidence and before final conclusions can be drawn.

In this context, another coincidence should be mentioned. It is an empirical fact [11.6] that $m_\mu/m_Z = 0.00115869(3)$ corresponds roughly to the Schwinger correction $\alpha/(2\pi)$. It is noteworthy that the number constant $\alpha_{geom}/(2\pi)$ [0.00115933] clearly agrees better with m_μ/m_Z than $\alpha/(2\pi)$ [0.00116141]. What this good numerical agreement to physics means is unclear.

The Fermi four-point interaction for radioactive decay leads to problems at high energies and breaks down, since the cross section diverges above an energy of ≈310 GeV, grows beyond all limits and the interaction probability becomes greater than 1. In the Standard Model, this problem is solved by the introduction of the $W^{+/-}$ bosons of mass m_W as virtual mediator particles. In addition, a finite interaction length $r_w = h_{bar}/(m_W c)$ is defined due to the Heisenberg uncertainty principle. It has a value of $\approx 2.5 \cdot 10^{-18}$ m and is today regarded as the smallest distance that can be observed experimentally.

Assuming that the Fermi theory is the low energy limit of the weak interaction of the Standard Model,

Formula 11.3 $g_w^2/(8 m_W^2) = G_F/2^{1/2}$

follows. By the introduction of $W^{+/-}$ bosons the interaction is no longer described by the coupling factor G_F, but analogous to the electric charge q_e a universal weak coupling constant g_w occurs. For low energies, that is to say in the case of small momentum transfers, the Fermi theory is still a good approximation.

If tabulated values for the Fermi constant $G_F\, q_e^2/(h_{bar}c)^3$ ($\approx 1.166 \cdot 10^{-5}$ GeV^{-2}) and for the mass $m_W c^2/q_e$ (≈ 80.385 GeV) are used, in analogy to the fine-structure constant α,

Formula 11.4 $g_w^2/4\pi \equiv \alpha_W \approx 1/29.5$

is obtained for the "weak charge" g_w, which characterizes the weak interaction via the coupling factor $G_F/2^{1/2}$ and the mass m_W.

11.5.1 The particle number density n and the mass m_Z

The rest energy that a virtual Z boson has as a mediator of the weak interaction amounts to (91.1876 ± 0.0021) GeV/c² according to CODATA. This energy is equivalent to a mass m_z of $\approx 1.625567 \cdot 10^{-25}$ kg. How is the Z-particle, corresponding to the photon in electrodynamics, able to localize this energy and to act as a virtual particle with mass m_z? Since the Z-particle interacts with the electron and the positron, its mass may possibly be associated with the particle number density n. Setting $\alpha = \alpha_{geom}$, $m_e = m_{e\_geom}$ and $kg = m^3/c^2$,

Relation 11.14 $\{m_z/eV\}\, n = \{m_z/eV\}\, c^{-13/4}\, h^{-1/4}\, 2^{35/4}\, \pi^{23/6}\, 10^{7/2} \approx 0.99996 \cdot (5/9)$

can be deduced by means of n from Table 5.1. The factor $10^{7/2}$ arises due to the relationship $\mu_0 \equiv 4\pi \cdot 10^{-7}$. What does Relation 11.14 mean? Does the mass $m_x = (9/5)\, m_z$ with a value of ≈164.145 GeV/c² possibly have an experimental or theoretical meaning?

Relation 11.14 does not seem to be a coincidence, because for the ionization energy of hydrogen IE_H (≈13.6 eV) and the ionization energy of helium IE_{He} (≈24.6 eV) a similar relation of the kind $IE_H \approx (5/9)\, IE_{He}$ applies.

In the Extended Standard Model of particle physics, the masses of the force particles are generated by an interaction via a relativistic scalar field or Higgs field, which has a nonzero vacuum expectation value (v) as a manifestation of the uncertainty principle. This scalar field is said to be omnipresent in the universe and it is assumed that the coupling to the Higgs field is proportional to the mass of the force particles. The connection between the mass of the W-particle and the Higgs field is mathematically represented by $m_W^2 = 4^{-1} g_w^2 v^2$. Substituting this expression in Formula 11.3, a relation of the form $v^2 = 1/(2^{1/2} G_F)$ results for the vacuum expectation value after a simple calculation. This expression allows to calculate the vacuum expectation value using the well-known Fermi constant G_F of ≈ 1.166364 GeV^{-2}. Remarkably,

Relation 11.15 $1/(2^{1/2} G_F) \approx (\mathbf{246.221}\ \text{GeV})^2 \approx (3/2 \cdot m_x c^2)^2 \approx (\mathbf{246.218}\ \text{GeV})^2$

applies. Because of Relation 11.15, the scalar field $v = (2^{1/2} G_F)^{-1/2}$ can thermodynamically be interpreted as the mean kinetic energy of a point particle with three degrees of freedom, which is in equilibrium with the "heat energy" $m_x c^2$. Because of $m_z = (5/9) m_x$, the mass of the Z-particle is based on thermodynamic interactions similar to the mass of the electron, and can be calculated by means of the number constants 2 and π and the constants h and c, since the "heat energy" $m_x c^2$ follows from the requirement $m_x c^2 n = 1$.

11.6 The collective magnetic interaction as a number constant and the associated energy scale

According to classical magnetostatics, the interaction energy of atomic dipoles, which are at a distance r from each other, is of the order of magnitude of $\mu_{Bohr}{}^2 \cdot \mu_0 / r^3$. The Bohr magneton μ_{Bohr} is defined by

Definition 11.3 $\mu_{Bohr} \equiv q_e\, h_{bar}/(2m_e) \approx 9.274 \cdot 10^{-24}$ A m²

and plays a significant role as a natural unit of the magnetic moment of the electron, since all magnetic moments of atoms are measured in units of the Bohr magneton. In magnetic solids magnetic dipole distances are typically ≈2 Å, so that the classical interaction energy or magnetic ordering temperature between atomic magnetic moments is on the order of 0.1–1 K. This means that dipole interactions between atomic magnetic moments are much too weak to account for effects where the observed magnetic ordering temperatures are orders of magnitude higher. For example, the paramagnetic Curie temperature Θ_P for Co is 1,415 K.

Classical dipole interactions cannot induce magnetism, and another collective interaction mechanism or magnetic exchange force must exist to achieve a nonvanishing long-range magnetization. Such an interaction must deeply be rooted in quantum mechanics, because the classical view, as explained earlier, provides energies only allowing interactions near the absolute temperature zero. Probably the exchange force is based on a combination of Coulomb interaction and Pauli principle. Today the opinion exists that the collective magnetism is related to superconductivity, crystallization or even the creation of masses.

11.6.1 The magnetic coupling constant of the long-range exchange interaction

With μ_{Bohr} (unit A m²) and the magnetic field constant $\mu_0 \equiv 1/(\varepsilon_0 c^2)$, the product $\mu_0\, \mu_{Bohr}{}^2$ can be formed with the unit J m³. Since the term $\mu_0\, \mu_{Bohr}{}^2$ has the same dimension as the Fermi constant G_F, a dimensionless "magnetic coupling constant" results in formal analogy to Definition 11.2. It shall be defined by

Definition 11.4 $\alpha_{Mag} \equiv (\mu_{Bohr}{}^2\, \mu_0\, 4^{-1})/(hc)/r_M{}^2$

which, if r_M is known, is a pure number and could be the mesoscopic origin of the collective ordering phenomenon magnetism. Is it also the reason why a matter volume is subdivided into magnetic domains? To get an interaction radius r_M with a physically "reasonable" value,

Hypothesis 11.2 $\alpha_M \equiv (\alpha_{grav}/\alpha_{geom})^{1/2} \approx 8.5\cdot10^{-19}$

shall axiomatically apply. Such a setting includes both electromagnetism (α) and gravity (α_{grav}) in the definition of α_M. Putting $\alpha_{Mag} = \alpha_M$, Hypothesis 11.2 can be rearranged to yield

Relation 11.16 $r_M = \alpha_{grav}^{-1/4}\, \alpha_{geom}^{1/4}\, (\mu_{Bohr}^2\, \mu_0\, 4^{-1})^{1/2}\, (hc)^{-1/2}$

for the interaction radius r_M which amounts to ≈12.675 µm using CODATA values for the natural constants. This corresponds roughly to the characteristic "magnetic" length L because of

Relation 11.17 $r_M/L \approx 0.9998$

Is this not a remarkable numerical coincidence? Using geometrized values, the length ratio r_M/L is uniquely determined on the basis of Hypothesis 11.2 and has a value of ≈1.001719. Does this number, mathematically represented by

Relation 11.18 $(r_M/L)_{geom} = (5/3)^{-1/4}\, 2^{-51/4}\, \pi^{47/6}$

have a physical meaning? The approximate length equality $r_M \approx L$ refers to a deep connection between the micro- and macroworld, because also the Newton constant G is involved. Are gravity, electricity and collective magnetism or magnetic pairing causes of a single field, and therefore all interconnected? All three interactions are long-range forces with the difference that gravitation acts between masses, that electricity is an interaction between charges and that magnetism represents a complex interplay between charge and motion. What is the driving force behind it? Is the characteristic length L the physical reason why in bulk materials magnetic domains form, and why electrons in large assemblies can reach a state of deepest energy by means of magnetic phenomena?

By means of the interaction radius $r_M \approx L$ a characteristic magnetic energy scale $h_{bar}cL^{-1}$ can be determined, which, due to Hypothesis 5.1, corresponds to $k_B\tau/(2\pi)$ or $\tau_{bar} \equiv \tau/(2\pi) \approx 180.5$ K, respectively.

Some ceramic materials do not show superconducting behavior just at very low temperatures, but at higher temperatures in the range of 100 K. The mechanism of the high-temperature superconductivity, that is, the current transport with almost no resistance, is controversial among experts, as is still unknown how the high critical transition temperatures come about. Magnetism or an antiferromagnetic interaction is an obvious explanation because in all cases strong magnetic fluctuations are observed. Something has to happen with the electrons to make the medium for the charge carriers, despite repulsive electrical interaction, even more permeable than in the best metallic conductor silver. In Table 11.1, experimental critical transition temperatures are given, which are interestingly related to the magnetic energy scale $h_{bar}cL^{-1}$ by simple fractions, pointing to a collective magnetism as the cause of the phenomena.

Table 11.1: Transition temperatures

Material	Reference	T_c (K)	T_c/τ_{bar}	
MgB_2		37	≈0.205	≈ 1/5
$Ba_{0.6}K_{0.4}Fe_2As_2$	[11.7]	38	≈0.210	≈ 1/5
$YBa_2Cu_3O_7$		90	≈0.499	≈ 1/2
$Tl_2Ba_2CaCu_2O_8$		108	≈0.598	≈ 3/5
$HgBa_2Ca_2Cu_3O_8$		135	≈0.748	≈ 3/4

Note: List of transition temperatures of some superconductors with particularly high transition temperatures T_C. As a comparison, normalized values obtained by division with the magnetic energy scale $\tau_{bar} = \tau/(2\pi) \approx 180.5$ K are listed in the last column.

Collective magnetism may also play a role in the Verwey transition of the stoichiometric magnetite Fe_3O_4. This material has a striking resistance behavior as a function of temperature. Single-crystalline material exhibits a sharp, first-order change in resistance at T_{Verwey} of ≈120 K by almost two orders of magnitude. At the transition, the material changes from a poor conductor (≈$2.5 \cdot 10^4$ μΩ cm) in the high-temperature phase (T > 120 K) to an insulator in the low-temperature phase (T < 120 K). In contrast to metals, in the case of magnetite, a first-order transition to worse electrical conduction occurs as the temperature is lowered. This insulator-semiconductor transition is highly pressure dependent and sensitive to changes in stoichiometry.

A. M. Bataille and coauthors [11.8] investigated the influence of thickness on the Verwey transition of magnetite epitaxially grown on α-Al_2O_3 without internal stresses. They found that layers with layer thicknesses larger than ≈20 nm show a clear, sharp Verwey transition at ≈120 K similar to massive samples, whereas layers with layer thicknesses smaller than ≈20 nm do not form such a transition. The observed characteristic thickness of ≈20 nm is on the order of the universal London length λ_{Lo} of ≈21 nm and underlines the relevance of this characteristic length for the Verwey transition. Therefore, it is not surprising that also the transition temperature T_{Verwey} close to ≈120 K correlates with the magnetic energy scale $h_{bar}c/L$. In units of τ_{quer}, the transition temperature T_{Verwey} is ≈0.665 or ≈2/3, respectively.

The magnetic properties of antiferromagnetic chromium are complex and an old subject of solid-state physics. At normal pressure, pure chromium undergoes a phase transition from a paramagnetic phase to a transversely polarized, antiferromagnetic ground state at the Neél temperature of ≈311 K, in which the magnetic moments are spatially modulated. This so-called spin density wave is a collective state with a broken translational symmetry and is not tightly coupled to the crystal lattice. Upon further reduction of the temperature, a transition to a longitudinally polarized spin density wave occurs at the spin-flip temperature $T_{spin\text{-}flip}$ of ≈121 K. Remarkably, similar to the Verwey transition, a value of ≈2/3 results for the ratio $T_{spin\text{-}flip}/\tau_{bar}$.

11.7 Representation of the anomalous magnetic moment of the electron by means of coupling constants

The anomalous magnetic moment of the electron a_e is a dimensionless measure that can stably be measured and that is very sensitive to interaction effects. When electrons are unbound, no binding corrections are necessary and the measurement can be performed with extremely high accuracy. This can be achieved by trapping individual electrons for days in a Penning trap by a clever arrangement of magnetic and electric fields, and spectroscopically investigating them at low temperature (4 K). The movement in the electromagnetic field is complicated and must be described quantum mechanically.

All energies are quantized, and the levels caused by the orbital motion (cyclotron angular frequency ω_c) are additionally split up due to the electron spin. This energy splitting (spin-flip angular frequency ω_s) is proportional to the g-factor of the electron g_e and the applied magnetic field. The ratio ω_s/ω_c amounts to $g_e/2$, when the magnetic field acting on the orbital and the spin movement is the same. For the local magnetic field the cyclotron angular frequency ω_c can be used.

The first accurate measurements of a_e in a Penning trap, which roughly corresponded to a cube with a side length of ≈3 mm, were made by R. S. Van Dyck and coworkers [11.9]. They obtained a value of 0.001159652400(200) by evaluating the experiments on the basis of a near Dirac point particle model. Today the experimental value of the dimensionless number a_e is stated as 0.00115965218085(76) with an accuracy of at least 12 decimal places after the decimal point [5.23].

In textbooks, the anomaly of the g-factor is explained by effects of QED. It is assumed that the effect is based on virtual, unobservable processes as outlined in Section 5.4.1. The calculations of the higher-order correction terms by means of QED are extremely complex and cannot be performed without the help of parallel computers. Although the first three terms can analytically be determined using algebra programs, no physics can be derived from them. The terms remain complex mathematical structures without any relation to physical reality.

The magnetic dipole associated with the electron spin is of fundamental importance in determining macroscopic properties. Why should not the electron's inherent tendency to long-range interactions also be observed in the measurement of a_e? If, in addition to the Schwinger contribution $\alpha/2\pi$, a term is introduced that describes the interaction of the electrons with each other due to their spin magnetic moments,

Relation 11.19 $a_{e\_theo} = \alpha_{geom}/(2\pi) + \{\alpha_M/(8\pi)\}^{1/3} \approx 0.001159652523$

could apply. If the impacts of two interactions are combined, a_{e_theo} coincides up to the ninth decimal place after the decimal point with experimental data. The magnetic coupling constant $\alpha_M \equiv \alpha_{grav}{}^{1/2}\alpha_{geom}{}^{-1/2}$ is a pure number independent of natural constants, which can be calculated by means of α_{geom} via Hypothesis 11.2 and Relation 11.6. In contrast to the anomalous magnetic moment of the muon a_μ, for a_e a third term can be found, which almost brings the deviation δa_e from the measured value to disappear. If, in Relation 11.19, the gravitational interaction term $\delta a_e = -\{4{\cdot}3/5{\cdot}\alpha_{grav}\}^{1/4}$ is added,

Ansatz 11.4 $a_{e\_theo} = \alpha_{geom}/(2\pi) + \{\alpha_M/(8\pi)\}^{1/3} - \{4{\cdot}3/5{\cdot}\alpha_{grav}\}^{1/4} \approx 0.0011596521884$

is obtained. This simple approach, compared to the sophisticated calculations of QED, is in agreement up to the 13th decimal place with the famous precision experiment of the Dyck–Schwinberg–Dehmelt group of the University of Washington [11.10], who stated 0.0011596521884(43) for a_{e_meas}. For 20 years, this value was the reference for a_e per se. Similar to the anomalous magnetic moment of the muon a_μ, only natural multiples of 2π occur in the expression a_{e_theo} for the calculation of the anomalous magnetic moment of the electron. And unlike the pure QED approach of today's doctrine, Ansatz 11.4 considers a gravitational interaction that must exist when an all-pervasive gravitational field is admitted.

Is the excellent agreement of a_{e_meas} with a_{e_theo}, which can be calculated solely with three dimensionless mathematical terms, a coincidence? In any case, Ansatz 11.4 is an astounding simplification to the complicated QED calculation that relies on particle exchange. It is therefore worthwhile to delve deeper into this coincidence, because each of the three number constants α_{geom}, α_M and α_{grav} is associated with a force field and originates independently of a_e in a different physical context.

References

[11.1] P. A. M. Dirac. A new basis for cosmology. Proceedings of the Royal Society of London A 165 (1938) 199–208

[11.2] K. Hirose, S. Saito and Yoshi. H. Ichikawa. Structure of period-2 step-1 accelerator island in area preserving maps. National Institute for Fusion Science (NIFS) 408 (1996)

[11.3] Wolfgang Pauli. Wissenschaftlicher Briefwechsel mit Bohr, Einstein, Heisenberg u.a. Band IV, Teil IV, A:1957, Springer Verlag Berlin

[11.4] H.J. Treder. Gravitation und weitreichende schwache Wechselwirkungen bei Neutrino-Feldern. Astronomische Nachrichten 295 (1974) 169–184

[11.5] Werner Heisenberg. Einführung in die einheitliche Feldtheorie der Elementarteilchen. S. Hirzel Verlag Stuttgart 1967

[11.6] S. I. Sukhoruchkin. The QED corrections in the Standard Model. Journal of Physics: Conference Series 171 (2009) 012064 [11 pages]

[11.7] M. Rotter, M. Tegel and D. Johrendt. Superconductivity at 38 K in the iron arsenide ($Ba_{1-x}K_x$) Fe_2As_2. arXiv: 0805.4630v2 [cond-mat.supr-con]
[11.8] A. M. Bataille, E. Vincent, S. Gota and M. Gautier-Soyer. Finite size effects in the Verwey transition of magnetite thin films. arXiv: cond-mat/0610291
[11.9] R. S. Van Dyck, Jr., P. B. Schwinberg, and H. G. Dehmelt. Precise measurements of axial, magnetron, cyclotron, and spin-cyclotron-beat frequencies on an isolated 1 meV electron. Physical Review Letters 38 (1977) 310–314
[11.10] R. S. Van Dyck, Jr., P. B. Schwinberg, and H. G. Dehmelt. New high precision comparison of electron and positron g-factors. Physical Review Letters 59 (1987) 26–29

12 Interpretations of astronomical measurements with universal parameters

12.1 The total mass density and the Hubble parameter

12.1.1 The mean total mass density

The mean mass density of visible matter, which consists of all the stars and hot gases of the observable universe, can be determined from the mass-luminosity relation. Assuming that matter is homogeneously distributed in space and that very large volumes are considered, the mass density of luminous matter is estimated at $\approx 10^{-28}$ kg m^{-3}.

By observing stars in spiral galaxies, it was discovered that there is not enough visible matter to explain the motions of baryonic matter using Newton's theory of gravitation. The dynamics can only be explained without altering physical laws by assuming additional "dark" or invisible matter that does not exchange energy with electromagnetic radiation in any spectral region. Galaxies obviously have to consist of something other than glowing stars, gases and microscopic solid particles accumulated as cosmic dust. From many observations and measurements, the astronomers estimated the "gravitating" total mass density ρ_{grav}, which includes both the known "luminous" matter and "dark" matter, at $1 \cdot 10^{-27}$–$5 \cdot 10^{-27}$ kg m^{-3}. This information is very uncertain and must be treated with caution, which makes it difficult to use the estimate as a benchmark. Nevertheless, a comparison is tried.

In thermodynamics, in a one-component, single-phase system, two state variables suffice to fully describe the state. This follows from the Gibbs phase rule which states the number of independent intensive variables for a system in equilibrium. For an ideal gas, temperature (T), pressure (P) and mass density (ρ) are related by

Formula 12.1 $P \sim \rho \, T$

All parameters in Formula 12.1 are intensive variables and do not depend on the size of the system. The material-dependent proportionality factor, called the specific gas constant (unit J K^{-1} kg^{-1}) or the individual gas constant, is the universal gas constant R_{gas} divided by the mean molar mass of a given gas. This constant is specific to a particular gas, while the universal gas constant is the same for any ideal gas. If for the creation and annihilation of matter a thermodynamic equilibrium is postulated in accordance with

Ansatz 12.1 $\text{rho}\ T(r = r_{rho}) \sim \rho\ T(r = r_e/2 \approx r_{pion\_charged})$

https://doi.org/10.1515/9783110612387-012

a density can be calculated from it. This picture suggests that in terms of thermodynamics the evolution of a closed system of many particles tends to a state of equilibrium, which can be described by a few parameters. For an adiabatic system at rest, such an equilibrium arises automatically due to the first law of thermodynamics. Assuming that there is a local thermodynamic equilibrium among the various particles and any statistical weights on either side of the proportionality of Ansatz 12.1 can be neglected, the mean mass density on very large scales is given by

Relation 12.1 $\rho \approx \mathrm{rho}\ T(r = r_{rho})\ T(r = r_e/2)^{-1} \approx 5.1 \cdot 10^{-27}\ \mathrm{kg\ m^{-3}}$

The value of $\approx 5.1 \cdot 10^{-27}$ kg m^{-3} obtained by means of Relation 12.1 corresponds to about three protons (more precisely ≈3.026) per cubic meter of space and is of the same order of magnitude as what astronomers determined on the largest scales that can still be detected experimentally. The rationale for the choice of the temperature $T(r = r_e/2 \approx r_{pion\_charged}) = m_e c^2/k_B$ is that this quantity has served well in the theoretical determination of the primordial abundance of ^{4}He and provided a result consistent with experimental facts. Since electromagnetic radiation (light) can oscillate independently in two directions and mass (sound) in three directions,[1] the mean “gravitating” total mass density ρ_{grav} is thus given by

Ansatz 12.2 $\rho_{grav} \equiv (2/3)\ \mathrm{rho}\ T(r = r_{rho}) T(r = r_e/2)^{-1} = (1/3)\ \mathrm{rho}\ r_e\ r_{rho}^{-1} \approx 3.4 \cdot 10^{-27}\ \mathrm{kg\ m^{-3}}$

In deriving Ansatz 12.2, the two temperatures $T(r = r_{rho})$ and $T(r = r_e/2)$ were replaced by the corresponding length standards $r_e/2$ (≈1.4 fm) and r_{rho} (≈123 nm). Ansatz 12.2 implies that the rate of creation of matter and the rate of annihilation of matter (radiation) in the universe are the same, that is, gravitational matter annihilates and regenerates from the vacuum energy rho similar to biological structures. Since matter particles are randomly generated and annihilated in the cosmic particle gas, on very large scales a stable total mass density is small compared to the mass density rho of $\approx 3.4 \cdot 10^{-27}$ kg m^{-3}, when the particles are independent of each other. This gravitational medium of the universe would be the reason why structures (galaxies) have formed. Interestingly, the assumption of a stationary universe, what Ansatz 12.2 represents, leads to a total mass density in the same order of magnitude estimated experimentally by astronomers today. In the past, proponents of a stationary view of the universe were independent of each other – Walther Hermann Nernst in Europe and W. D. McMillan in America.

1 The physical cause of this significant difference is unknown.

12.1.2 The radiation temperature T_{rad}

The special theory of relativity of Albert Einstein allows the conversion of the mass density ρ_{grav} into the relativistic energy density $\rho_{grav}\,c^2$ ($\approx 3\cdot 10^{-10}$ J m^{-3}), and thus the calculation of a radiation temperature T_{rad} via the radiation Formula 10.5 of Max Planck. Both matter and radiation are only special forms of energy. It results in

Definition 12.1 $\quad T_{rad} \equiv (\rho_{grav}\,c^2)^{1/4}\,2^{-3/4}\,\pi^{-5/4}\,15^{1/4}\,k_B^{-1}\,(hc)^{3/4} \approx 25.2\text{ K}$

By Definition 12.1 relativistic energy (mass) is completely transformed (conservation of energy) into radiation, which can escape unhindered into space. The transformation of matter into pure energy and vice versa is an experimental fact and the basis for the description of the interaction between electron, positron and photon. Is $k_B T_{rad}$ the energy in which the cosmic dust, the solid of interstellar matter per se, is embedded?

12.1.3 The Hubble parameter H_0

It is expected that Newton's constant G depends on few universal properties of the universe. The dimension of the Newton constant (unit m^3 s^{-2} kg^{-1}) is a combination of a mass density (unit kg/m^3) and the square of an inverse characteristic time. Using the cosmological Hubble parameter H_0 (unit s^{-1}) for the inverse time and ρ_{grav} for the mass density, the correlation of ρ_{grav} with another measurement can be checked. Simple dimensional analysis yields

Ansatz 12.3 $\quad G \sim H_0^2\,\rho_{grav}^{-1}$

The currently most accurate value [12.1] of the Hubble parameter H_0, as determined by observations with the Hubble Space Telescope, is (74.2 ± 3.6) km s^{-1} Mpc^{-1} in traditional units or $\approx 2.4\cdot 10^{-18}$ s^{-1} in SI units. The Hubble parameter is not a natural constant, but a proportionality factor of the approximate linear relationship ($c\,z \approx H_0\,D$) between the relative distances (D) of "near" astronomical objects (D > 1 Mpc) and the redshifts (z < 1) measured from their spectra. According to current doctrine, the Hubble relationship is interpreted in such a way that all objects move away from us as long as there is no longer any gravitational attraction. But basically, the profound nature of the Hubble law, that is, the correlation of redshift with distance, is not fully understood. The main problem in determining H_0 is that the proportionality factor must be determined on far-off objects to eliminate local effects. This is the reason why the Hubble parameter is not yet known in a higher precision today, because for the measurement of distances physical assumptions that are subject to relatively large uncertainties must be made.

Setting $H_0 \approx 2.405 \cdot 10^{-18}\ s^{-1}$, as determined from observations with the Hubble telescope, and using a value of $\approx 3.4 \cdot 10^{-27}$ kg m^{-3} for the gravitational total mass density ρ_{grav}, as calculated by Ansatz 12.2,

Relation 12.2 $\quad H_0^2\, \rho_{grav}^{-1} \approx 1.022 \cdot 8\pi\, G$

holds. Can from this numerical coincidence be concluded that

Hypothesis 12.1 $\quad G\, \rho_{grav}\, H_0^{-2} = (8\pi)^{-1}$ with $\rho_{grav} = (1/3)\, \text{rho}\, r_e\, r_{rho}^{-1}$

applies without uncertainties in H_0? Is the Hubble parameter H_0 determined by ρ_{grav} and the Newton constant G, that is, the speculative Hypothesis 12.1? Or is it simply a remarkable coincidence that the "repulsive form of gravitation" (antigravity) can be described by Hypothesis 12.1?

It can only then be decided whether in Hypothesis 12.1 the proportionality factor is in fact 8π, if the Hubble parameter H_0 is experimentally determined more accurately. Recently, the same group [12.2] lowered the uncertainty to 2.3% and came up with a value of (73.48 ± 1.66) km s^{-1} Mpc^{-1} that perfectly fits in Hypothesis 12.1. Namely, with **≈73.44 km** s^{-1} **Mpc**$^{-1}$ ($\approx 2.380 \cdot 10^{-18}\ s^{-1}$) for H_0 and geometrized values for the other constants, Hypothesis 12.1 is exactly fulfilled. It is noteworthy that the calculation value inferred from Hypothesis 12.1 lies decisively within the range of the experimental Hubble parameter given by Adam G. Riess and coauthors. With the definition $\rho_{krit} \equiv 3H_0^2/(8\pi G)$ of the Friedmann equations, Hypothesis 12.1 leads to the relationships $\rho_{grav} = \rho_{krit}/3$ and $\rho_{krit} = \text{rho}\, r_e\, r_{rho}^{-1}$.

12.1.4 The Jeans length: an interesting identity

By means of dimensional considerations, a length defined by

Definition 12.2 $\quad \lambda^2 = k_B T\, m^{-1}\, G^{-1}\, \rho^{-1}$

can be determined. This length scale is also known as the Jeans length [12.3], which defines a limit of a gravitating gas between stability and instability. According to this theory, all scales larger than the Jeans length are unstable with respect to a gravitational collapse, and smaller scales are stable. If λ is redefined according to

Relation 12.3 $\quad \lambda_{Jeans}^2 \equiv k_B T(r = r_{rho})\, m_e^{-1}\, G^{-1}\, \rho_{grav}^{-1}$

some algebraic calculations yield

Relation 12.4 $\quad \lambda_{Jeans} = (3/2)^{1/2}\, L_{large}$

12.1.5 The quantization of redshift: a controversial topic

In 1976, William G. Tifft [12.4] investigated the internal motions of the Milky Way, the Andromeda galaxy, the Vortex galaxy and the galaxy NGC 2903 in the constellation Leo. He observed that redshift values within individual galaxies are not uniform, but occur only in specific discrete values of a multiple of 70–75 km s^{-1}. A year later [12.5], he extended his research to galaxies in pairs and groups, thus extending his observations on gravitational interactions to larger-scale structures.

Many of his observations indicated that redshift differentials (Δcz) between pairs of galaxies take on preferred values that are various multiples of a basic interval. From the largest line splitting in the double galaxy NGC 4922, for example, a value of ≈73 km s^{-1} can be calculated for the basic interval if the observed line splitting equals five states of the basic interval. Redshift data from several galaxies of the entire Coma galaxy cluster best fit a basic interval of ≈72.5 km s^{-1} when all redshifts are corrected for the Earth orbital motion and galactic rotation (300 km s^{-1}).

In the third publication [12.6], W. G. Tifft completed his empirical quantization concept by the observation that from the measurements of McLaughlin, made in Nova Herculis from about December 20, 1934, until about January 1, 1935, stable velocities can be deduced, which differ by an internal offset of ≈72.5 km s^{-1} or a multiple thereof.

The observation that the redshift does not change continuously but is a discrete variable triggered a lot of additional research after the first three papers of W. G. Tifft. Extending the measurements to galaxies scattered all over the sky, however, makes the analysis more demanding as galactocentric corrections must be applied. W. G. Tifft and other independent researchers found out that even half or one-third of the basic interval of ≈72 km s^{-1} can occur as periods.

Unfortunately, the physical effect has become increasingly unbelievable, since insufficient accuracy of the measurements cannot be ruled out for ever shorter periods. The peculiar quantization of the differential redshift data in situations where the differences of velocities are small and the measurement accuracy plays a minor role seems to be an experimentally verified fact. Nevertheless, the problem is rarely addressed today because the phenomenon does not fit into the existing concept of the cosmological paradigm. For most astronomers, the observation of W. G. Tifft is the result of systematic observation effects in combination with a too small amount of data.

The statement by W. G. Tifft is based on many measurements of relative redshifts between pairs of galaxies. For our galaxy and the neighbor galaxy Andromeda, which is still visible to the naked eye, the concept of W. G. Tifft, if the unexplained effect is universal, should also apply. The current distance ($R_{Andromeda}$) between our galaxy and Andromeda is ≈780 kpc [12.7]. Cepheids found in the Andromeda Nebula allow this distance to be estimated quite well by systematic observation. If the experimentally well-proven, basic value δv_{Tifft} of ≈72.5 km s^{-1} is divided by the value H_0 calculated from the Hubble Hypothesis 12.1 and the length $R_{horizon}$ (≈873 kpc),

Relation 12.5 $\delta v_{Tifft}/(H_0 R_{horizon}) \approx R_{horizon}/R_{Andromeda}$

is obtained. The term $H_0 \cdot R_{horizon}$ represents a velocity (≈64.1 km s^{-1}) in relation to the Hubble time H_0^{-1} and the particle horizon $R_{horizon}$. Does Relation 12.5 possibly provide an explanation for why W. G. Tifft and others observed a grouping of velocities of ≈72.5 km s^{-1}?

12.2 Interpretation of two experimental findings of cosmic radiation by means of coupling constants

From the cosmos, we constantly receive high-energy particles that are not from the Earth's crust, as the intensity increases with increasing altitude. This particle shower, called cosmic radiation, consists of atomic nuclei with an elemental abundance which, with some exceptions, roughly corresponds to the elemental abundance of the sun. Where the cosmic, high-energy particle shower originates is not clearly known. It is believed today that star explosions could be the main sources. The energy of cosmic radiation is in any case millions of times greater than the energy that can be generated in the laboratory by the LHC at CERN, where nuclei are accelerated to energies that are "only" a hundred to a thousand times larger than their rest energies.

12.2.1 E. Regener: the permanent ionizing primary radiation

From measurements [12.8] of the ionization strength as a function of altitude, E. Regener estimated the energy flux at $\approx 3.53 \cdot 10^{-3}$ erg cm^{-2} s^{-1}, which is dissipated in a vertical column of air per area and time by ionization at the "boundary" of the atmosphere. The measurements of the discharges in air were carried out with highly isolated, static electrometers with an automatic registration in balloons up to an altitude of ≈27 km.

By multiplication with the factor 4/c, the measured energy flux density of $\approx 3.53 \cdot 10^{-3}$ erg cm^{-2} s^{-1} ($\approx 3.53 \cdot 10^{-6}$ J m^{-2} s^{-1}) can be converted into a spectral energy density (unit J m^{-3}) of a high-energy, electromagnetic primary radiation consisting of x-ray and gamma quanta. This conversion results in a value of $\approx 4.71 \cdot 10^{-14}$ J m^{-3}. What is the source of this permanent ionizing primary radiation of extraterrestrial origin?

If the vacuum energy density (rho c^2) and a combination of electromagnetism (α) and radioactivity (α_F) is responsible for the permanent ionization measured by E. Regener,

Ansatz 12.4 $(\varepsilon/V)_{Regener} = 3/5\, \alpha_F\, \alpha\, \mathrm{rho}\, c^2$

could be a possible explanation. Using CODATA values, $(\varepsilon/V)_{Regener}$ is given by

Relation 12.6 $(\varepsilon/V)_{Regener} \approx 4.7 \cdot 10^{-14}$ J m$^{-3} \approx 0.29$ eV cm^{-3}

which agrees quite well with the value of $\approx 4.7 \cdot 10^{-14}$ J m^{-3} measured by E. Regener with an ionization chamber.[2]

Ampere's law of magnetic flux states that charge currents are the sources of magnetic fields. This means that the permanent ionizing primary radiation could be related to magnetic fields. Assuming a uniform distribution of ionizing and magnetic energy densities, according to

Formula 12.2 $(\varepsilon/V)_{magnetic} = B^2/(2\mu_0) = (\varepsilon/V)_{Regener} \approx 4.7 \cdot 10^{-14}$ J m^{-3}

a mean magnetic flux density equivalent to the energy density $(\varepsilon/V)_{Regener}$ can be calculated. Using Formula 12.2, a value of $\approx 3.4 \cdot 10^{-10}$ T or $\approx$ 3.4 µG is obtained. Interestingly, this value is on the order of magnitude of the global magnetic flux density of 2–8 µG in our galaxy.

12.2.2 The cosmic ray energy spectrum

Upon impact of cosmic radiation, which in addition to electrons and positrons consists for the most part of high-energy protons and helium nuclei (alpha particles), collisions occur in the earth's atmosphere with atomic nuclei of the air. This results in a cascade of secondary particles or a so-called particle shower, which is mainly composed of pions and muons. The cosmic ray energy spectrum, that is, the number of cosmic particles per energy interval, area, time and solid angle interval as a function of their energy (ε), extends over orders of magnitude of $\approx 10^3$ to $\approx 10^{20}$ eV and drops very steeply with increasing energy. Below $\approx 10^{15}$ eV the particle flux follows a power law proportional to $\varepsilon^{-2.7}$, and above this the spectrum becomes steeper and obeys a power law proportional to ε^{-3}. The place in the spectrum where the steepness changes is called the "knee" of the energy spectrum. At this point physics seems to change.

The highest particle energies observed [12.9] lie above $\approx 10^{20}$ eV. This place in the energy spectrum is referred to as the "ankle" of the spectrum. The highest particle energy ever measured so far is $\approx 3.2 \cdot 10^{20}$ eV [12.10]. At such high energies, the flux is only about one particle per square kilometer in 200 years. How are the particles accelerated to such high energies? What is the cause of the "knee" in the energy spectrum at ε_{knee}? Why are particles with energies greater than $\approx 5 \cdot 10^{19}$ eV observed despite the Greisen–Zatsepin–Kuz'min (GZK) cutoff [12.11]? What is the highest energy ε_{max} that can occur at all? These are exciting physical questions.

2 Interestingly, this value is close to the energy density of the cosmic microwave background radiation ($\approx$2.7 K) of $4.2 \cdot 10^{-14}$ J m^{-3} ($\approx$0.26 eV cm^{-3}).

Choosing

Ansatz 12.5 $\varepsilon_{knee} = \alpha_{grav}\, \alpha_F\, hc\, L_2(R_{horizon})^{-1}$

for the energy threshold ε_{knee}, which is still not understood and discussed controversially, a numerical value of $\approx 4.3 \cdot 10^{15}$ eV or 4.3 PeV for ε_{knee} is obtained. In the work of M. Amenomori and coauthors [12.12], the differential all-particle cosmic ray flux in the range from 10^{14} to 10^{17} eV is shown, as measured by the Tibet-III air-shower array. As a comparison, additional experimental data of other antenna fields are shown in this illustration. From this, according to the authors, an energy threshold of $\approx 4 \cdot 10^{15}$ eV can be deduced for the kink in the energy spectrum. Andreas Haungs [12.13] gives a value of (3.2 ± 1.2) PeV for the position of the knee determined from several experiments.

The length $L_2(R_{horizon})$ used in Ansatz 12.5 is unimaginably small and amounts to $\approx 4 \cdot 10^{-70}$ m, that is, it is still much tinier than the already unreasonable small Planck length. As a result, the energy equivalent $hc/L_2(R_{horizon})$ corresponding to the length $L_2(R_{horizon})$ is very large and can certainly never be mimicked in the laboratory. Is $L_2(R_{horizon})$ the space length that cannot be further subdivided, that is, the fundamental limit beyond which physical theories can no longer make any statements?

In order to reach the earth, relativistic particles with a threshold impulse (p) of $\approx 3.2 \cdot 10^{20}$ eV c^{-1} must be confined by a magnetic field (B), which determines the maximum achievable momentum perpendicular to the magnetic field lines. It must apply

Formula 12.3 $r_{Larmor} = p/(q_e B)$

For the Larmor radius r_{Larmor} the radius of the particle horizon can be used according to Relation 6.8, which approximately amounts to half of the intergalactic length L_{large}. According to Formula 12.3, the magnetic flux density preventing the escape of a particle from the particle horizon with a threshold impulse of $\approx 3.2 \cdot 10^{20}$ eV c^{-1} is then $\approx 4 \cdot 10^{-11}$ T or ≈ 0.4 µG, respectively. Kim and coauthors [12.14] indeed reported a magnetic flux density of 0.3–0.6 µG over an intergalactic range of 1–2 Mpc.

The interpretation of experimental findings of cosmic radiation by means of universal quantities may be coincidental. It is noteworthy that simple relationships can be found using physical quantities of other technical fields without the need for additional parameters in modeling. The exact mechanism of the acceleration of particles to such extremely high energies is certainly not solved. Perhaps, Ansatz 12.5 may provide an indication of which interaction processes might be responsible. The smallness of the length $L_2(R_{horizon})$ is also a big mystery. There is no relation to our reality for this length, since it is in a technically never realizable order of magnitude.

References

[12.1] Adam G. Riess, Lucas Macri, Stefano Casertano et al. A redetermination of the Hubble constant with the Hubble Space Telescope from a differential distance ladder. The Astrophysical Journal 699 (2009) 539–563

[12.2] Adam G. Riess, Stefano Casertano, Wenlong Yuan et al. New parallaxes of galactic cepheids from spatially scanning the Hubble Space Telescope: Implications for the Hubble constant. https://arxiv.org/pdf/1801.01120.pdf, queried on March 10, 2018

[12.3] J. H. Jeans. The stability of a spherical nebula. Philosophical Transactions of the Royal Society (London) 199 (1902) 1–53

[12.4] William G. Tifft. Discrete states of redshift and galaxy dynamics I: Internal motions in single galaxies. The Astrophysical Journal 206 (1976) 38–56

[12.5] William G. Tifft. Discrete states of redshift and galaxy dynamics II: Systems of galaxies. The Astrophysical Journal 211 (1977) 31–46

[12.6] William G. Tifft. Discrete states of redshift and galaxy dynamics III: Abnormal galaxies and stars. The Astrophysical Journal 211 (1977) 377–391

[12.7] T. J. Cox and Abraham Loeb. The collision between the Milky Way and Andromeda. Monthly Notices of the Royal Astronomical Society 386 (2008) 461–474

[12.8] E. Regener. Der Energiestrom der Ultrastrahlung. Zeitschrift für Physik 82 (1933) 666–669

[12.9] Kenneth Greisen. End to the cosmic-ray spectrum? Physical Review Letters 16 (1966) 748–750.

[12.10] High resolution fly's eye experiment. University of Utah. http://www.cosmic-ray.org, queried on October 9, 2011

[12.11] M. Takeda, N. Hayashida, K. Honda et al. Extension of the cosmic-ray energy spectrum beyond the predicted. Greisen-Zatsepin-Kuz'min cutoff. http://arxiv.org/PS_cache/astro-ph/pdf/9807/9807193v1.pdf, queried on October 16, 2011

[12.12] M. Amenomori, S. Ayabe, X. J. Bi et al. The energy spectrum of all-particle cosmic rays around the knee region. observed with the Tibet-III air-shower array. http://arxiv.org/PS_cache/arxiv/pdf/0803/0803.1005v1.pdf, queried on October 16, 2011

[12.13] Andreas Haungs. Energy spectrum and mass composition around the knee by EAS measurements. http://arxiv.org/PS_cache/astro-ph/pdf/0212/0212481v1.pdf, queried on October 16, 2011

[12.14] K. T. Kim, P. P. Kronberg, G. Giovannini and T. Venturi. Discovery of intergalactic radio emission in the Coma-A1367 supercluster. Nature 341 (1989) 720–723

13 Our star – the sun

> We do not even understand how our own star really works.
> (*William A. Fowler*)[1]

The sun governs our solar system in terms of mass and energy production and is the only star that can be studied from not too far away. But despite this fact, many features of the sun are not clear.

13.1 The spectral radiant flux density: the transition between disturbed and quiet solar activity

The radiation of the sun can be subdivided into the always existing radiation of the "quiet sun" with little or no solar activity and the radiation of the "disturbed sun" originating from rapidly varying flare emission. Optical telescopes provide a spectrum of the spectral radiant flux density that follows the Planck curve of a black body radiator with a temperature of ≈ 5,800 K. Observations with radio telescopes [13.1] measure spectral radiant flux densities above a wavelength greater than ≈1 cm, which are larger than those that formally correspond to a temperature of ≈5,800 K.

The meaning of "quiet sun" or "disturbed sun" is not clearly defined, and the effect why within certain wavelength ranges the radiation of the sun is enhanced is not understood. It is known that the radiation at different wavelengths originates in different regions of the solar atmosphere. The photosphere emits radiation at wavelengths of millimeters, while longer wavelength radiation has its origin in the chromosphere or corona. It is interesting that at the critical wavelength of ≈1 cm the radio sun has approximately the same radius as the photosphere and the brightness appears evenly over the entire surface. For wavelengths in the meter range, the sun looks larger than the optical disk. The subdivision of the solar flux density spectrum into two curves at the critical wavelength λ_{crit_sun} (≈ 1 cm) marks a critical point where the difference of the radio radiation between the "quiet" and the "disturbed" sun ceases to exist.

From a mass and a temperature an expression according to

Relation 13.1 $c/\lambda_{crit\_sun} \sim q_e^{-2}\,(4\pi\varepsilon_0)\,m^{-1/2}\,(k_BT)^{3/2}$

can be formed, which has the dimension of a length. If the electron mass m_e is chosen for the mass m and, according to Definition 12.1, the temperature T_{rad} of ≈25.2 K for

1 Cauldrons in the Cosmos, Claus E. Rolf and William S. Rodney, University of Chicago Press (1988).

https://doi.org/10.1515/9783110612387-013

the temperature T, a value of ≈1.02 cm (≈29.4 GHz) results for the length λ_{crit_sun} by means of Relation 13.1. It is interesting that, using dimensional considerations by means of the electron mass and the energy $k_B T_{rad}$, a wavelength in the order of the critical radio wavelength can be calculated, where the solar radio radiations of the quiet and the disturbed sun are equal and are transformed into a black body radiation of ≈ 5,800 K.

The fact that an energy on the order of magnitude of the Debye temperature (≈ 25 K) of solid helium could play a role is not illogical, since the surface of the sun consists of ≈91 at% of atomic hydrogen and ≈8.9 at% of helium. Is this agreement an amazing coincidence, or is Relation 13.1 an indication of a possible, certainly unconventional, explanation of the complicated issue of solar activity? Can it be that $k_B T_{rad}$ is effective both in the solid state and in the plasma? This would correspond to an idea of J. D. Van der Waals, who argued that long-range interactions are acting beyond phase boundaries.

According to Benz [13.2], the flux density of the "quiet" sun is approximately 1,862 sfu (1 sfu = 10^4 Jy = 10^{-22} W m^{-2} Hz^{-1}) at the measurement wavelength of ≈1 cm. This value is based on measurements of various telescopes around the globe made during the sunspot years 1964 and 1976. From this value, using geometric considerations, an effective temperature of ≈10^4 K or ≈0.862 eV can be calculated at ≈1 cm in the center of the optical disk [13.2]. Interestingly, this thermal radiation being constant in time corresponds to about $3T(r = a_B) = 3 \cdot k_B^{-1} m_0 c^2$ or ≈ 9963 K (≈0.858 eV), respectively.

13.2 Interpretation of the critical frequency of the ionosphere by means of universal parameters

The ionosphere plays an important role in the propagation of electromagnetic radio waves around the globe because it reflects specific wavelengths. Radio waves are used for communication over long distances and, radiated vertically into the atmosphere, can penetrate the ionosphere only above a critical frequency f_0F_2 of typically 2–15 MHz. Lower frequencies are totally reflected by charged particles and can no longer leave the earth. The critical frequency, where complete reflection occurs, is dependent on the angle of incidence, correlates with solar activity and shows differences between day and night as well as seasonal variations. Critical frequencies for normal incidence are continuously measured by ionosondes distributed over the globe and their values are recorded.

Figure 13.1 shows as a typical example the critical frequencies measured in Juliusruh in the years 1957–2013 and the solar sunspot numbers in the same time period. The illustration clearly manifests the seasonal fluctuations and the range of values of the ionospheric frequency f_0F_2.

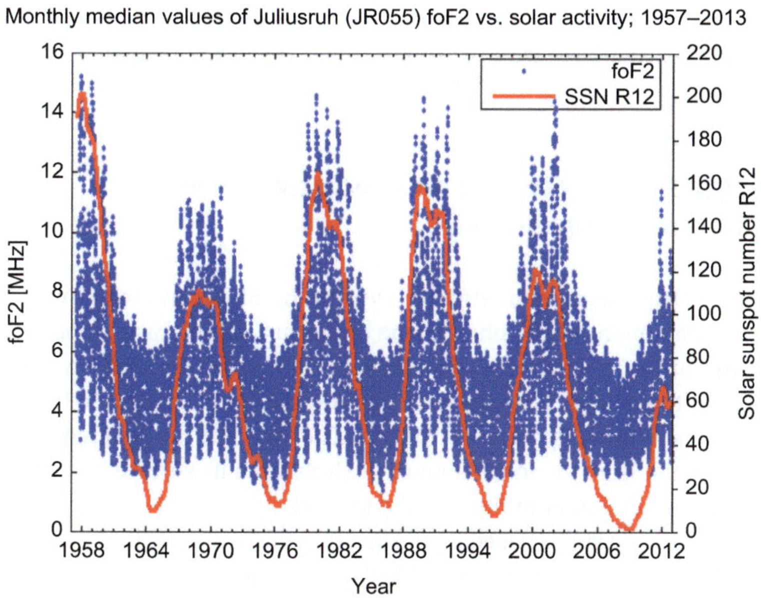

Figure 13.1 Monthly median values of the critical frequency f_0F_2.

Note: Survey of the monthly median values of the critical frequency f_0F_2, which were measured in Juliusruh by means of pulse echo sounding during the years 1957–2013. As a comparison, the relative sunspot numbers are plotted in red, showing that the sun and the ionosphere are related systems. Which processes are responsible for this correlation? Source: Monthly Survey of Ionospheric Data from Juliusruh [13.3] with kind permission of the Leibniz-Institute of Atmospheric Physics.

Low frequencies imply small photon energies that concern extremely low-energy transitions, such as hyperfine splittings or quantized rotations of molecules. The energy $h \cdot (f_0F_2)$ of a photon of frequency f_0F_2 of 10 MHz corresponds to a temperature $(h/k_B) \cdot (f_0F_2)$ of ≈0.5 mK, which scales, according to Definition 6.2, fairly well with the smallest fundamental temperature T_{Min} of ≈0.8 mK. Assuming that the neutrinos of the sun are responsible for the temporal fluctuations of the critical frequency f_0F_2, and the neutrino mass scales with the mass m_0,

Relation 13.2 $$f_0F_2 \sim q_e^{-2} (4\pi\varepsilon_0)\, m_0^{-1/2} (k_B T_{Min})^{3/2} \approx 7.64 \text{ MHz } (\approx 39.3 \text{ m})$$

can be deduced in analogy to Relation 13.1 and by simple dimensional arguments. Relation 13.2 is dependent only on natural constants.

If instead of m_0 the cut-off mass m_ν calculated by Definition 6.6 is used for the neutrino mass, a critical frequency of ≈5.18 MHz or a radio wavelength of ≈57.9 m results. Is it not astonishing that the monthly median values obtained from measurements of the ionosonde of Juliusruh during several years are oscillating with a period of ≈11 years around the calculated value of ≈ 5MHz? Is this just a coincidence based on dimensional analysis?

13.3 The electric sun hypothesis: an alternative view of the internal energy source

The model of the electric sun is an alternative view to the Standard Model, which cannot or hardly explain many of the phenomena of the sun. In any case, nobody is able to look into the sun or to take measurements in it. The model that the temperature in the interior of the sun must be several million Kelvin to maintain the fusion of hydrogen to helium running is pure speculation. The fact that the observed line spectra of the sun are not from the inside, but exclusively from its surface, also puts the prediction about the solar interior not on solid ground.

The model of the electric sun is based on the fact that the processes in the photosphere with its cellular structure are comparable with electric arc effects. From this it is deduced that the sun is a luminous, positively charged electrode (anode) that supports an electric plasma. The cathode is the surrounding space (heliosphere) and exists only virtually. The nature of the solar interior plays a minor role, since the electrical discharge is supplied by an external galactic circuit. In the interior an incompressible nuclear matter of homogeneous density without an internal energy source is assumed. All electrons are separated from nuclear matter and are free to move. The positively charged nuclear matter carries the mass responsible for gravitation and at the same time prevents gravitational compression due to a repulsive Coulomb field.

Assuming that within a cylinder of area $\sigma_{Thomson}$ and length λ_{crit_sun} there is exactly one free electron before a photon is elastically scattered at the electrons, the number density of electrons n_{e_sun} can be estimated. The cross-section for Thomson scattering $\sigma_{Thomson}$ can be derived by classical arguments and is $(8\pi)\ 3^{-1}\ r_e^2$ or $6.65 \cdot 10^{-29}\ m^2$. In Thomson scattering, photon energy is absorbed by a charged particle regardless of the frequency and is emitted again as electromagnetic radiation at the same frequency. Using

Ansatz 13.1 $\sigma_{Thomson}\ n_{e\_sun}\ \lambda_{crit\_sun} = 1$

a value of $\approx 1.5 \cdot 10^{30}\ m^{-3}$ is obtained for the number density n_{e_sun}.

If the sun is a plasma of electrons and charged particles of mass m_{grav}, their density can be estimated by $n_{e\_sun} m_{grav}$, which is ≈2332 kg m^{-3} when Relation 6.21 and CODATA values for the natural constants are used for the calculation of m_{grav}. From the radius and the mass of the sun, a mean density ρ_{sun} of 1,408 kg m^{-3} can be calculated. With $n_{e\_sun} m_{grav}$ and ρ_{sun}, a density ratio according to

Relation 13.3 $\rho_{sun}/(n_{e\_sun}m_{grav})$

can then be determined, which amounts to ≈0.604.

Remarkably, this ratio corresponds to the space filling factor of a simple hexagonal primitive lattice of contacting rigid spheres. The density of such a sphere packing is in fact $\pi \cdot 3^{-3/2}$ or ≈0.605. Is a simple hexagonal primitive sphere packing the optimal structure of positively charged particles of mass m_{grav} that repel each other and want to be as far apart as possible? Is the sun a macroscopic, homogeneous quantum phase of crystalline order with m_{grav} as a fundamental building block in a lake of unbound electrons of the number density n_{e_sun}? Where and how are the elements made?

References

[13.1] John D. Kraus. Radio astronomy. McGraw-Hill Book Company, New York, 1966, p 329
[13.2] A. O. Benz. Quiet and slowly varying radio emissions of the sun. Landolt-Boernstein LB VI/4B 4.1.1.6 (2009)
[13.3] Monthly survey of ionosphere data from Juliusruh. http://www.ionosonde.iap-kborn.de/mon_main.htm, queried on October 19, 2013

14 Phenomenological cataloging of particles with Hall fractions

> The fractional quantum Hall effect reveals that ostensibly indivisible quanta – in this case the electron charge q_e – can be broken into pieces through self-organization of phases. The fundamental things, in other words, are not necessarily fundamental.
> (*Robert B. Laughlin*)[1]

A big, unsolved problem of today's particle physics is the explanation of the particle masses observed. Even after decades of research, no accepted theory is able to compute the observed mass pattern from scratch within the experimental error limits. By the Brout–Englert–Higgs mechanism, no mass values can be derived either, because the theory does not explain or solve the problem, but only shifts it. At least the theory contributes something to its understanding, unfortunately without clarifying the nature of gravitation.[2]

Due to our ignorance of the physical phenomena that occur on a scale of ≈1 fm, particle masses, such as the mass of the muon, the proton or the neutron, appear in today's physical models as free parameters that are only experimentally accessible and allow no theoretical calculation from scratch. The inability to make predictions is due to the fact that the process of renormalization of quantum field theory, which today is the theoretical basis of particle physics or the Standard Model, prevents any access to the computation of free parameters for reasons of principle. The renormalization of quantum field theory was introduced without any physical reason to avoid mathematical infinites.

In addition, the mass of the electron is considered by to current doctrine as a fundamental observable and requires no explanation from scratch because of the mathematical trick of renormalization. The effective (measured) electron mass is composed of a free mass of the underlying theory and an infinite number of photon interactions, which can be calculated perturbatively by means of quantum electrodynamics (QED) due to the weak coupling to the electromagnetic field. Although this approach to explaining and describing experimental data is extremely successful, it is mathematically not self-consistent and therefore unsatisfactory. Physically disturbing is the fact that for the free electron mass as well as for the free charge infinite values must be assumed. Such a theory cannot be the ultimate truth; too great are the mathematical and physical inadequacies.

In Section 6.1, a completely different view is taken and it is assumed that the electron mass is not a free parameter, but can be defined axiomatically as a function

1 A Different Universe, Basic Books (2005), p. 77.

2 As long as the origin of mass is not clarified, all mass values of particle physics must be considered as "natural constants".

https://doi.org/10.1515/9783110612387-014

of c and h (or k_B), which are freely selectable. The axiomatic determination of the fine structure constant by the number constant α_{geom} effectuates, by means of the likewise freely selectable magnetic field constant μ_0, that the charge is also not a free parameter that must be measured. These definitions (hypotheses) based on thermodynamic concepts do not produce any singularities with the requirement that infinite values have to be assigned to both the free electron mass and the free charge.

In this chapter, we will investigate whether the hypotheses of previous chapters lead to verifiable predictions of masses, magnetic moments, charge radii and partial lifetimes of the particle zoo [14.1]. If this can be accomplished, then various physical phenomena can be related to a few fundamental assumptions that are not themselves derived from particle properties. The book thus becomes logically consistent by means of additional experimental findings.

In order to be able to interpret the diversity of the experimental data, additional numbers are necessary. To limit the selection, it is useful to look for numbers that have already played a central role in other areas of physics and have been experimentally verified. The fractions observed in the fractional quantum Hall effect might be suitable. This "number zoo" is universal and has been an experimental fact for many years. Like the existence of the muon, the fractional quantum Hall effect that occurs with utmost accuracy and reproducibility has not been predicted by any theory. The effect was discovered by chance when looking at low temperatures for the crystallization of electrons.

In two-dimensional charge systems, the Hall resistance R_{Hall} does not change continuously at low temperatures (<4 K) as a function of high magnetic field (several Tesla), but in steps according to the simple formula given as follows:

Formula 14.1 $R_{Hall}^{-1} = \upsilon\,(q_e^2/h)$

It is fascinating that, regardless of the temperature, the geometry or the purity of the sample, only natural numbers or specific fractions are observed for the filling factor $\upsilon \equiv p/q$.

In the article by H. L. Stormer et al. [14.2], a typical graph is printed, from which for υ the values

4 3 2 5/3 7/5 4/3 1 4/5 5/7 2/3 3/5 4/7 5/9 4/9 3/7 2/5 1/3

can be gained. In other experiments, the fractions

7/2 8/3 5/2 7/3 8/5 3/2 9/7 7/9 1/2 3/8 2/7 2/9 1/5 1/7

were also observed when only natural numbers from the set {1,2...9} are allowed for p and q. The restriction for p and q was applied because all fractions are thus expressed by the number 1 or the primes 2, 3, 5 or 7, which are related to the decomposition of the reciprocal volume of the nine-dimensional hypersphere according to Relation 3.5.

In the physics of the fractional quantum Hall effect, even-numbered fractions are extremely puzzling, since quantum states or phases of matter that can be characterized by such fractions are very rarely observed. This is the reason, for example, all fractions with $q = 4$ are missing in the aforementioned enumeration.

Today, the observed fractions are explained by assuming that in two-dimensional charge systems, bound states are formed that consist of an electron to which a given number of magnetic flux quanta are attached.

14.1 The muon-electron mass ratio

Since its discovery in the investigation of cosmic radiation in 1936, it is still unclear why the muon exists at all. No theory has ever predicted its existence. In contrast to the electron, the particle does not play a recognized role in explaining the structure of matter until now. Why does the muon exist and weigh about 207 times more than the electron? Why is it in all other properties like the electron except for its mass and its stability? The Standard Model has no answers to these questions.

The mass of the muon of ≈105.7 MeV is on the order of magnitude of the energy $L_2(L)^{-1}\ h_{bar}c/q_e$ of ≈108.3 MeV. The length scale $L_2(L)$ was already used in Section 8.1 for the explanation of the nucleon density and it is therefore not so wrong to assume that this length scale could also occur at the muon. But what is the reason for the energy difference of ≈2.6 MeV compared to the measured muon mass?

Since muons behave like heavy, unstable electrons, it is not so far-fetched that what we observe as a muon could be an emergent phenomenon of the energy $L_2(L)^{-1}h_{bar}c$, the rest energy m_ec^2 and some kind of binding interaction. The classical electron radius (≈2.8 fm) corresponds to the range of nuclear forces and suggests that in fact electrons might latently play a role in the nucleus. The latent presence of electrons in the nucleus is also supported by the fact that after the disintegration of muons almost only electrons remain. Excited atoms also emit light, although no photons are contained in the atom itself.

The most accurately measured parameter of the muon is its magnetic moment. In spite of excellent agreement between experiment and present theory, there is an unexplained, small difference for which no satisfactory explanation has yet been found. The observed discrepancy between the experimental data and the present modeling, which is based entirely on QED, is interpreted by certain physicists as a hint that the current QED model of the muon is incomplete due to a possible substructure.

As an alternative view to the QED model, another idea is presented in Section 11.4, which describes the anomalous magnetic moment of the muon a_μ solely using the number constants α_{geom} and α_F. Interestingly, the phenomenological Ansatz 11.3 with the additional correction $\delta a_\mu \sim \alpha_F/\alpha_{geom}$ results in a very good agreement with the experimental value. In this view, it seems that not only α_{geom} and α_F play a role, but also some kind of mixing α_F/α_{geom} must be included to characterize the anomalous magnetic moment of the muon.

Some kind of mixing $\alpha_{grav}/\alpha_{geom}$ analogous to the mixing α_F/α_{geom} also appears in Section 11.6.1. By means of Definition 11.4, the ratio $\alpha_{grav}/\alpha_{geom}$ gives an astonishing relation between the length $L \sim \tau^{-1}$ and the magnetic moment of the electron $\mu_e = 1/2\ \mu_B\, g$, if the spin g-factor is classically set to 1 and not to the measured value of ≈ -2.0023. The mixing $\alpha_{grav}/\alpha_{geom}$ also decisively improves the difference between the measured value of a_e and that calculated by Relation 11.19.

By analogy with Definition 11.4,

Definition 14.1 $\alpha_{mag} \equiv (\mu^2\, \mu_o\, 4^{-1})/(hc)/r_{mag}^2$ with $\mu = h_{bar} q_e/2m_\mu$

shall apply. Setting $\varepsilon_{mag} = h_{bar}c/r_{mag}$,

Definition 14.2 $\varepsilon_{mag} \equiv 4\, m_\mu\, \alpha_{mag}^{1/2}\, (hc/\mu_o)^{1/2}\, c/q_e$

follows, which is equivalent to

Relation 14.1 $\varepsilon_{mag} = f_\mu\, \alpha_{mag}^{1/2}$

if ε_{mag} is expressed in units of m_ec^2/q_e ($\approx$0.511 MeV) and

Definition 14.3 $f_\mu = 4\, m_\mu/m_e\, (2\alpha)^{-1/2}$

is used for f_μ. Both f_μ and α_{mag} are dimensionless numbers.

According to P. A. M. Dirac, the factor $(2\alpha)^{-1} \approx 137/2$ occurring in f_μ corresponds to the ratio q_m/q_e of magnetic charge to electric charge. To date, however, no matter has been found that carries a magnetic charge, and it is completely unclear why the symmetry between electric and magnetic quantities that is inherent in classical field theory is not experimentally observed, although the symmetry of Maxwell's equations demands the existence of magnetic monopoles.

Setting in Relation 14.1 $\alpha_{mag} = \alpha_F/\alpha_{geom}$ and using CODATA values for all other parameters, a value of $\approx$0.672 MeV is obtained for ε_{mag}. Surprisingly, according to Section 8.2, this value corresponds to the standard deviation ($\approx$0.670 MeV) of the mean nuclear binding energy per nucleon ($\approx$8.03 MeV) of the 2,932 isotopes compiled by Thayer Watkins.

Assuming that the muon is a bound state and that the energy quantities $L_2(L)^{-1}h_{bar}c$ and m_ec^2 are its building blocks, which are held together by a binding energy proportional to ε_{mag}, then the muon-electron mass ratio can be approximated by

Ansatz 14.1 $m_\mu/m_e = \lambda_{e\_bar}/L_2(L) - 7 - 1/3\, f_\mu(\alpha_F/\alpha)^{1/2}$

Care was taken that according to the principle of simplicity, only integers or simple Hall fractions occur. Using geometrized values, Ansatz 14.1 yields for $m_\mu/m_e \approx 206.72$, which is quite close to the tabulated CODATA value (2,014) of 206.7682826(46). If, in addition, the correction terms $+2/5\ f_\mu \alpha_F^{1/2}$ and $+3/5\ f_\mu(\alpha_F\alpha)^{1/2}$ are added, that is, Ansatz 14.1 is extended to

Ansatz 14.2 $$m_\mu/m_e = \lambda_{e\_bar}/L_2(L) - 7 - 1/3\ f_\mu(\alpha_F/\alpha)^{1/2} + 2/5\ f_\mu\ \alpha_F^{1/2} + 3/5\ f_\mu(\alpha_F\alpha)^{1/2}$$

a geometrized value of ≈**206.768072646** can be calculated for m_μ/m_e.

The calculation of m_μ/m_e can simply be accomplished because the factor f_μ is proportional to the mass ratio, and Ansatz 14.2 therefore represents a simple linear equation for the arithmetic of m_μ/m_e. Further terms would improve the deviation from the tabulated CODATA value, but at the same time destroy the symmetry between (α_F/α_{geom}), (α_F) and $(\alpha_F\alpha_{geom})$. Unless it is clear what the underlying physical origin of Ansatz 14.2 is and how much the tabulated CODATA value of m_μ/m_e depends on QED models, it makes no sense to add an additional term.

All binding terms are without doubt of magnetic origin and seem to hold together different energy quanta. There are both attractive and repulsive binding terms. Magnetic forces must be important in the creation of masses, because not only the electron, the muon and the proton, but even the uncharged neutron have magnetic moments. The three fractions 1/3, 2/5 and 3/5 that occur in Ansatz 14.2 are common Hall fractions, which also appear in the graph of H. L. Stormer et al. [14.2].

If, in the following, geometrized values for α, m_e and G are used in a calculation, then the geometrized value 206.768072646 according to Ansatz 14.2 is also employed for m_μ/m_e. This ensures that no experimental data, that is, only the number constants 2 and π, are utilized in the computation. Of course, geometrization also effectuates the ratio $\lambda_{e\_bar}/L_2(L)$ to depend only on 2 and π. A simple calculation yields the elegant expression $\mathbf{2^{16}}\ \pi^{-5}$ for it.

14.2 Mass ratios of other particles

The provocative concept of calculating the muon-electron mass ratio by Ansatz 14.2 provides, with only three binding terms, a very good approximation to the experiment, but it lacks a physical basis for a deeper understanding. The question therefore arises as to whether a similar concept might be applicable to other mass ratios as well, which agree with the experimental data within their error limits. If it is true that mass ratios can be represented with simple Hall fractions, force constants and few energy quanta as building blocks, this does not prove the picture to be right, but is possibly a hint to reflect it in greater depth.

To calculate the mass of the muon, it was assumed that the fundamental energy $L_2(L)^{-1}\, h_{bar}c$ cannot exist as matter alone and manifests itself in the real world solely as a composite system together with electrons. Could this also be correct for the fundamental energies $r_{grav}^{-1}\, h_{bar}c$ and $L_2(\Lambda)^{-1}\, h_{bar}c$? According to Section 8.2, the energy $L_2(\Lambda)^{-1}\, h_{bar}c$ seems to be responsible for the strong interaction or nuclear force, which holds together protons and neutrons in the nucleus of an atom. If this is true, the energy $L_2(\Lambda)^{-1}\, h_{bar}c$ should not play a role in describing the mass of the electron, the muon and the tauon.

In fact, it can easily be verified that

Ansatz 14.3 $$m_{pion\_neutral}/m_e = 1/7\ \lambda_{e\_bar}/r_{grav} + \lambda_{e\_bar}/L_2(\Lambda) - 9 + 2/9\ f_\mu \alpha_F^{1/2} \approx 264.1421$$

with a single magnetic binding term already coincides with the experimental value of 264.1426(12) within the error bounds.

Due to experimental inaccuracy, it does not matter if for the mass ratio m_μ/m_e the tabulated (206.768283) or the geometrized value (206.768072) according to Ansatz 14.2 is used in the calculation of f_μ. Ansatz 14.3 contains the fractions 1/7 and 2/9, which are numbers of the quantum Hall effect.

Similarly, the mass ratio $m_{pion\_charged}/m_e$ can be approximated. Using

Ansatz 14.4 $$m_{pion\_charged}/m_e = 1/7\ \lambda_{e\_bar}/r_{grav} + \lambda_{e\_bar}/L_2(\Lambda) + 3/2\ f_\mu(\alpha_F \alpha)^{1/2}$$

for $m_{pion\_charged}/m_e$, a value of ≈273.13157 is obtained, which agrees with the tabulated value of 273.13204(68) within the error limits. The occurring fraction of 3/2 is a rare even-numbered Hall fraction, experimentally observed only in recent years.

In addition to the mass ratios of the muon or the pions, many other mass ratios of particle physics and even atomic physics can be calculated by means of number constants and Hall fractions. The formalism guarantees that the structure of the particles is related, thus ensuring interactions between the particles. Thus, all ratios are independent of measured quantities and man-made natural constants, and only geometrized values must be used in the calculation of the mass ratios.

Like the length ratio $\lambda_{e\_bar}/L_2(L)$, the length ratios λ_{e_bar}/r_{grav} and $\lambda_{e\_bar}/L_2(\Lambda)$ depend exclusively on 2 and π when geometrized values are utilized.

While the ratio $\lambda_{e\_bar}/L_2(\Lambda)$ can be derived, according to Definition 5.2, by multiplying the ratio $\lambda_{e\_bar}/L_2(L)$ by $2^4\alpha_{geom}$, the ratio λ_{e_bar}/r_{grav} corresponds trivially to the mass ratio m_{grav}/m_e, already derived in Section 6.5.1.

The ratio $\lambda_{e\_bar}/L_2(\Lambda)$ is equivalent to $\mathbf{2^{14}}\ \pi^{-17/3}$, whereas for the ratio λ_{e_bar}/r_{grav}, according to Relation 6.24, the term $\mathbf{2^{-3}}\ \pi^{25/3}$ results, which can be decomposed into the product $2^3\,\pi^9\,\alpha_{geom}$. Together with Relation 9.29, it follows that λ_{e_bar}/r_{grav} is equal to L/a_H multiplied by the fine-structure constant α_{geom}. Is there a physical interpretation for the latter?

14.2.1 Compilation of mass ratios

In the following, all particles discussed so far and a multitude of others are compiled without text. There is no deep mathematical theory behind it, and it is pure phenomenology aiming to uncover the regularities hidden in the data similar to the concept of quantum numbers in particle physics (baryon number, lepton number, strangeness, charm, etc.) that were introduced as an aid for the classification of particle reactions.

Neutral pion

$m/m_e = 264.1426(12)$
$= 1/7\, \lambda_{e\_bar}/r_{grav} + \lambda_{e\_bar}/L_2(\Lambda) - 9 + 2/9\, f_\mu(\alpha_F)^{1/2} \approx 264.1421$

Charged pion

$m/m_e = 273.13205(68)$
$= 1/7\, \lambda_{e\_bar}/r_{grav} + \lambda_{e\_bar}/L_2(\Lambda) + 3/2\, f_\mu(\alpha_F\alpha)^{1/2} \approx 273.13157$

Charged kaon

$m/m_e = 966.102(31)$
$= 1/2\, \lambda_{e\_bar}/r_{grav} + 4\, \lambda_{e\_bar}/L_2(\Lambda) - 7/4\, f_\mu(\alpha_F/\alpha)^{1/2} \approx 966.086$

Neutral kaon

$m/m_e = 973.806(47)$
$= 1/2\, \lambda_{e\_bar}/r_{grav} + 4\, \lambda_{e\_bar}/L_2(\Lambda) + 7 - 6/5\, f_\mu(\alpha_F/\alpha)^{1/2} \approx 973.810$

Eta

$m/m_e = 1{,}072.14(4)$
$= 3/5\, \lambda_{e\_bar}/r_{grav} + \lambda_{e\_bar}/L_2(\Lambda) + 6 - 4/5\, f_\mu(\alpha_F/\alpha)^{1/2} \approx 1{,}072.17$

Rho-meson resonance

$m/m_e = 1{,}517.1(5)$ [775.26 ± 0.25 MeV]
$= 6/7\, \lambda_{e\_bar}/r_{grav} + \lambda_{e\_bar}/L_2(\Lambda) + 4 - 3/5\, f_\mu(\alpha_F/\alpha)^{1/2} \approx 1{,}517.1$

$= 1{,}505(2)$ [769.0 ± 1 MeV]
$= 6/7\, \lambda_{e\_bar}/r_{grav} + \lambda_{e\_bar}/L_2(\Lambda) - 9 \approx 1{,}505$

Omega-meson resonance

m/m_e = 1,531.6(2) [782.65 ± 0.12 MeV]
= $6/7\ \lambda_{e\_bar}/r_{grav} + 2\ \lambda_{e\_bar}/L_2(\Lambda) - 6 - f_\mu(\alpha_F/\alpha)^{1/2} \approx 1{,}531.5$

Proton

m/m_e = 1,836.15267389(17) CODATA value 2014
= 1,836.15267245(75) CODATA value 2010
= $\lambda_{e\_bar}/r_{grav} + 4\ \lambda_{e\_bar}/L_2(\Lambda) - 3/5\ f_\mu(\alpha_F/\alpha)^{1/2} + f_\mu(\alpha_F\alpha^2)^{1/2} + 9/5\ f_\mu(\alpha_F\alpha^3)^{1/2}$
≈ 1,836.15267169 geometrized value for m_μ/m_e
≈ 1,836.15267088 CODATA value (206.768284) for m_μ/m_e

Neutron

m/m_e = 1,838.68366158(90) CODATA value 2014
= 1,838.6836605(11) CODATA value 2010
= $\lambda_{e\_bar}/r_{grav} + 4\ \lambda_{e\_bar}/L_2(\Lambda) + 2 - 1/5\ f_\mu(\alpha_F/\alpha)^{1/2} + 5/9\ f_\mu(\alpha_F\alpha)^{1/2} + f_\mu(\alpha_F\alpha^3)^{1/2}$
≈ 1,838.68365691

Eta prime meson

m/m_e = 1,874.33(12)
= $\lambda_{e\_bar}/r_{grav} + 6\ \lambda_{e\_bar}/L_2(\Lambda) - 9 - 8/3\ f_\mu(\alpha_F/\alpha)^{1/2} \approx 1{,}874.35$

Phi-meson resonance

m/m_e = 1,995.04(4) [1,019.461 ± 0.019 MeV]
= $6/5\ \lambda_{e\_bar}/r_{grav} - 4\ \lambda_{e\_bar}/L_2(\Lambda) + 9 + f_\mu(\alpha_F/\alpha)^{1/2} \approx 1{,}995.00$

Lambda

m/m_e = 2,183.337(12)
= $6/5\ \lambda_{e\_bar}/r_{grav} + 4\ \lambda_{e\_bar}/L_2(\Lambda) - 4/5\ f_\mu(\alpha_F/\alpha)^{1/2} + 2/9\ f_\mu(\alpha_F)^{1/2} \approx 2{,}183.334$

Sigma plus (Σ^+)

m/m_e = 2,327.54(14)
= $7/5\ \lambda_{e\_bar}/r_{grav} - 4\ \lambda_{e\_bar}/L_2(\Lambda) - 5 + 1/3\ f_\mu(\alpha_F/\alpha)^{1/2} \approx 2{,}327.54$

Neutral sigma (Σ^0)

$m/m_e = 2{,}333.942(47)$
$= 7/5\ \lambda_{e\_bar}/r_{grav} - 4\ \lambda_{e\_bar}/L_2(\Lambda) + 7/5\ f_\mu(\alpha_F/\alpha)^{1/2} \approx 2{,}333.949$

Sigma minus (Σ^-)

$m/m_e = 2{,}343.349(59)$
$= 7/5\ \lambda_{e\_bar}/r_{grav} - 4\ \lambda_{e\_bar}/L_2(\Lambda) + 9 + 5/3\ f_\mu(\alpha_F/\alpha)^{1/2} \approx 2{,}343.300$

Neutral xi

$m/m_e = 2{,}573.12(39)$
$= 7/5\ \lambda_{e\_bar}/r_{grav} + 6\ \lambda_{e\_bar}/L_2(\Lambda) - 7 - 6/5\ f_\mu(\alpha_F/\alpha)^{1/2} \approx 2{,}573.12$

Charged xi

$m/m_e = 2{,}586.52(14)$
$= 7/5\ \lambda_{e\_bar}/r_{grav} + 6\ \lambda_{e\_bar}/L_2(\Lambda) + 6 - f_\mu(\alpha_F/\alpha)^{1/2} \approx 2{,}586.39$

Omega minus

$m/m_e = 3{,}272.90(57)$
$= 2\ \lambda_{e\_bar}/r_{grav} - 8\ \lambda_{e\_bar}/L_2(\Lambda) - 3 + f_\mu(\alpha_F/\alpha)^{1/2} \approx 3{,}272.85$

Tauon

$m/m_e = 3{,}477.15(31)$
$= 2\ \lambda_{e\_bar}/r_{grav} + 3 \approx 3{,}477.21$

Neutral D meson (D^0)

$m/m_e = 3{,}649.40(10)$
$= 2\ \lambda_{e\_bar}/r_{grav} + 7\ \lambda_{e\_bar}/L_2(\Lambda) + 2 - 6/5\ f_\mu(\alpha_F/\alpha)^{1/2} \approx 3{,}649.34$

Charged D meson ($D^\pm$)

$m/m_e = 3{,}658.74(18)$
$= 2\ \lambda_{e\_bar}/r_{grav} + 7\ \lambda_{e\_bar}/L_2(\Lambda) + 9 + 2/3\ f_\mu(\alpha_F/\alpha)^{1/2} \approx 3{,}658.80$
$m(D^\pm) - m(D^0) \approx 4.83\ \text{MeV}$ [PDG : 4.77(8) MeV]

Deuteron

m/m_e = 3,670.48296785(13) CODATA value 2014
= 3,670.4829652(15) CODATA value 2010
$= 2\,\lambda_{e\_bar}/r_{grav} + 8\,\lambda_{e\_bar}/L_2(\Lambda) - 2 - 6/5\, f_\mu(\alpha_F/\alpha)^{1/2} + 3/2\, f_\mu(\alpha_F)^{1/2}$
$+ f_\mu(\alpha_F\alpha)^{1/2} + 8/5\, f_\mu(\alpha_F\alpha^2)^{1/2} + 4/3\, f_\mu(\alpha_F\alpha^3)^{1/2} \approx 3670.48296561$

Charged D_S meson ($D_s^\pm$)

m/m_e = 3,851.87(22)
$= 11/5\,\lambda_{e\_bar}/r_{grav} + \lambda_{e\_bar}/L_2(\Lambda) + 6 - 3/5\, f_\mu(\alpha_F/\alpha)^{1/2} \approx 3{,}851.80$
$m(D_s^\pm) - m(D^\pm) \approx 98.62$ MeV [PDG : 98.69(5) MeV]

Lambda c baryon plus (Λ_c^+)

m/m_e = 4,474.5(3)
$= 13/5\,\lambda_{e\_bar}/r_{grav} - 2\,\lambda_{e\_bar}/L_2(\Lambda) + 8 \approx 4{,}474.49$

Neutral sigma c baryon (Σ_c^0)

m/m_e = 4,801.9(3)
$= 11/4\,\lambda_{e\_bar}/r_{grav} + \lambda_{e\_bar}/L_2(\Lambda) + 1 - 4/5\, f_\mu(\alpha_F/\alpha)^{1/2} \approx 4{,}801.94$
$m(\Sigma_c^0) - m(\Lambda_c^+) \approx 167.297$ MeV [PDG : 167.290(17) MeV]

Sigma c baryon plus (Σ_c^{++})

m/m_e = 4,802.3(3)
$= 11/4\,\lambda_{e\_bar}/r_{grav} + \lambda_{e\_bar}/L_2(\Lambda) + 2 - 5/4\, f_\mu(\alpha_F/\alpha)^{1/2} \approx 4{,}802.35$
$m(\Sigma_c^{++}) - m(\Lambda_c^+) \approx 167.505$ MeV [PDG : 167.510(17) MeV]

J/psi

m/m_e = 6,060.514(22)
$= 7/2\,\lambda_{e\_bar}/r_{grav} - \lambda_{e\_bar}/L_2(\Lambda) + 5 + 3/5\, f_\mu(\alpha_F/\alpha)^{1/2} - 3/2\, f_\mu(\alpha_F)^{1/2} \approx 6{,}060.523$

Neutral B meson

m/m_e = 10,331.9(3)
$= 6\,\lambda_{e\_bar}/r_{grav} - 4\,\lambda_{e\_bar}/L_2(\Lambda) + 9 \approx 10{,}331.8$

The preceding compilation indicates a commonality of combinable number constants, that is, mass ratios can be calculated as sums of a few dimensionless terms in combination with natural numbers and Hall fractions. The extraordinarily close correspondences with experimental data – mostly within their error limits – are astounding and should be an incentive to experimentally determine even more precisely the particle mass ratios, which have been well known for many years. Perhaps the occurring fractions in combination with the quantum numbers of the decay processes help reveal hidden regularities or even clarify the inner mechanism of matter creation that is so intensively sought today.

The lengths r_{grav} and $L_2(\Lambda)$ determine to 95% the masses of the hadrons. They are therefore the fundamental quantities of matter creation in the perspective presented in this chapter. Recently, researchers at the Nuclear Research Centre CERN established the existence of a resonance state at (4,449.8 ± 1.7 ± 2.5) MeV with a width of (39 ± 5 ± 19) MeV, which they called a "pentaquark-charmonium" state [14.3]. Whether this "exotic" hadron, which actually was not on the agenda of the LHCb collaboration, is a pentaquark or some other dynamic artifact, it is a real signal without doubt. It is astonishing that this state can also be interpreted very simply within its error limits as an emergence of the lengths r_{grav} and $L_2(\Lambda)$ as well as the number 5. Namely, the following applies:

$$m/m_e = 8{,}708(6)$$
$$= 5\,\lambda_{e\_bar}/r_{grav} + \lambda_{e\_bar}/L_2(\Lambda) \approx 8{,}710 \ (\approx 4{,}451\ \text{MeV})$$

For the sigma-hyperon, the mass of the positively charged particle $m(\Sigma^+)$ significantly differs from the mass of the negatively charged particle $m(\Sigma^-)$. The Particle Data Group [14.1] compiles (8.08 ± 0.08) MeV for $m(\Sigma^-) - m(\Sigma^+)$. The calculated difference is ≈15.76 electron masses or ≈8.05 MeV, if CODATA values for the conversion are used. This mass difference surprisingly corresponds to the average nuclear binding energy per nucleon (≈8.03 MeV) of the 2,932 isotopes compiled by Thayer Watkins.

14.3 The transformation of the muon

Not only the masses, but also the partial lifetimes, that is, the mean times required to transform particles into other matter particles, cannot be calculated from the ground up by the Standard Model, without using experimental data and continuously introducing new parameters.

In analogy to the theory of emission of light quanta from excited atoms via the usual radiation process, Enrico Fermi successfully described the decay of particles by means of a four-point interaction. The Fermi theory in its main outlines is still valid today and provides quantitative expressions for the mean lifetime, as well as the shape of the electron emission spectrum. In the Fermi theory, the total number of electrons and neutrinos is not constant. They are created at the emission and are annihilated when they are absorbed.

The mean lifetime τ of a particle transforming into three massless other particles is given by [14.5]

Formula 14.2 $\tau = 192\pi^3\, G^{\star-2}\, M^{-5}\, h_{bar}{}^7\, c^{-4}$

All influences of the transformation are absorbed in the dynamic parameter G* with the unit Jm^3 and the mass M. The kinematic factor $192\pi^3$ is called phase space factor. It is a measure of the motion in space and is as important for the decay as the interaction strength G* and the mass M.

Each measured lifetime is associated with the unit time. To convert measured time values into dimensionless numbers, a reference time is necessary. Much like m_e serves as a reference for the mass, it is straightforward to use the quantity $h/(m_e c^2)$ as a reference for the time. When converting a measured time by dividing it with the conversion factor $h/(m_e c^2)$, always CODATA values for h, m_e and c must be employed. The reference time $h/(m_e c^2)$ is about $8.1 \cdot 10^{-21}$ s. In the following, mainly decay processes with partial lifetimes greater than this reference value are investigated.

Except for the neutron, all particles disintegrate in different ways. In almost only one way the charged pion (99.99%), the xi minus particle (99.9%) or the muon (98.6%) disintegrate into other material particles. Since single-channel decays are very rare, for all other transformations, the measured total lifetime τ_{tot} must be corrected by its branching ratio B_i to obtain the mean partial decay time $\tau_i = \tau_{tot}/B_i$ of the decay channel i.

With the dimensionless quantities $x_1 = \tau_{tot}\,(m_e c^2/h)_{CODATA}$ and $x_2 = B_i$, Formula 14.2 yields

Relation 14.2 $(x_1/x_2)\, h/(m_e c^2) = 192\pi^3\, G^{\star-2}\, M^{-5}\, h_{bar}{}^7\, c^{-4}$

which allows by means of the mass M and the experimental quantities τ_{tot} and B_i to calculate a "Fermi constant" G* that is representative of the decay channel i.

In analogy to Definition 11.2, the coupling factor G * can then be related to a length r_{ww} via a dimensionless coupling constant α_{ww}. This leads to

Definition 14.4 $\alpha_{ww} = G^\star/(hc)/r_{ww}{}^2$

From this, the interaction energy $\varepsilon_{ww} = h_{bar}c/r_{ww}$ can be calculated by means of G* and α_{ww}, if it is known. Using Relation 14.2 and Definition 14.4, after some algebraic manipulations, the energy ε_{ww} in units of $m_e c^2/q_e$ may be written as

Definition 14.5 $\varepsilon_{ww} \equiv 24^{-1/4}\, \alpha_{ww}{}^{1/2}\, (M/m_e)^{5/4}\, (x_1/x_2)^{1/4}$

Because $\varepsilon_{ww} \sim (x_1/x_2)^{1/4}$, the experimental relative error $\Delta\varepsilon_{ww}/\varepsilon_{ww}$ is given by

Relation 14.3 $\Delta\varepsilon_{ww}/\varepsilon_{ww} = 1/4\, |\Delta x_1/x_1| + 1/4\, |\Delta x_2/x_2|$

if M/m_e is assumed as an error-free parameter. In analogy to Relation 14.1, only dimensionless numbers occur in Definition 14.5.

Since there are no experimental data for M/m_e and α_{ww}, the energy ε_{ww} is a function of these two parameters, which for each decay channel must be searched by numerical testing. For α_{ww}, however, at least an estimate is possible by using experimental, single-channel lifetime ratios. For example, in analogy to Section 11.3, from the experimental lifetime ratio $\tau_{pion\_charged}/\tau_{muon} \approx \alpha_{muon}/\alpha_F$, a value of $\approx 1.6\ \alpha_F\alpha$ for the dimensionless coupling α_{muon} can be gained. To limit the diversity, it makes sense for the mass M to use only fundamental energy quanta, which are important in building particle masses.

The muon has an average lifetime of 2.1969811(22) μs and decays in 98.6(4) percent of the cases into an electron and, according to the current doctrine, into an antielectron neutrino and a muon neutrino. In the Standard Model, the Fermi constant G_F ($\approx 1.166 \cdot 10^{-5}$ GeV^{-2}) is calculated from the measured lifetime using Formula 14.2 with $M = m_\mu$ and additional radiative corrections. On the other hand, the coupling factor G^* amounts to approximately $1.184 \cdot 10^{-5}$ GeV^{-2} using the simple Relation 14.2 without any corrections. Due to Relation 14.3, the experimental relative error $\Delta\varepsilon_{ww}/\varepsilon_{ww}$ is about 0.1%.

Setting $M = m_\mu$, $\alpha_{ww} = 1/4\ \alpha_{muon}$ ($\approx 0.4\ \alpha_F\alpha$), and using CODATA values for the constants, a value of ≈0.66 MeV is obtained for ε_{ww}. This surprisingly corresponds roughly to the magnetic energy $\varepsilon_{mag} = f_\mu(\alpha_F/\alpha)^{1/2}$, that is, the energy scale, which plays a role in the calculation of the muon mass by means of Ansatz 14.1. Does this correspondence suggest that the muon forms a resonance state with the energy $f_\mu(\alpha_F/\alpha)^{1/2}$, which is responsible, on the one hand, for muon's stability and, on the other hand, its disintegration into an electron, an antielectron neutrino and a muon neutrino?

Using geometrized values and converting the measured mean life to a dimensionless quantity, the following facts were evaluated by numerical testing:

Mass ratio M/m_e	$\mathbf{m_\mu/m_e - 2}$	
Branching ratio	98.6(4)%	
Coupling constant α_{ww}	$\mathbf{1/4\ (7/4 - \alpha)\ \alpha_F\alpha}$	
Resonance energy ε	$\mathbf{f_\mu(\alpha_F/\alpha)^{1/2}}$	
Deviation $\lvert\varepsilon_{ww}/\varepsilon - 1\rvert$	0.03%	[98.70% yields 0.001%][3]
Experimental relative error $\Delta\varepsilon_{ww}/\varepsilon_{ww}$	0.1%	

The fraction 7/4 occurring in the coupling constant α_{ww} appears in several places in this book. However, this number is not an observed Hall fraction for reasons already mentioned.

Since the branching ratio of the radiative decay has a large experimental error, the SM value of 1.30% shall be used to search for a suitable conversion process. Numerical testing gives:

3 The SM prediction [14.4] for the branching ratio of the radiative decay is 1.30%.

Mass ratio M/m_e	**2**
Branching ratio	1.30%
Coupling constant α_{ww}	$\mathbf{1/4\ \{9/8 - 3/4\ (\alpha/\pi)\}\ \alpha_F\alpha^2}$
Resonance energy ε	$\mathbf{f_\mu(\alpha_F\alpha^2)^{1/2}}$
Deviation $\lvert\varepsilon_{ww}/\varepsilon - 1\rvert$	0.1%

The parameter α/π is used in quantum electrodynamics by theoreticians to describe radiation processes. If the disintegration of the muon proceeds exclusively on two channels according to the two processes listed earlier, which are highlighted in bold, the two branching ratios and the lifetime are fixed. Using

Relation 14.4 $\tau_{tot}/B_i = 24\ (\alpha_{ww})_i^{-2}\ (M/m_e)_i^{-5}\ \varepsilon_i^4\ h/(m_ec^2)$

which can be derived from Definition 14.1 by simple algebraic transformations, and the requirement $B_1+B_2 = 1$, a mean lifetime of 2.1969826 μs [exp. 2.1969811(22) μs] and the decay probabilities 0.987054 or 0.012946 can be calculated. It is noteworthy that such a simple model gives such an excellent match with the measured lifetime of the muon. It goes without saying that geometrized values and CODATA values for $h/(m_ec^2)$ are mandatory. If for m_μ/m_e, the CODATA value is used in the calculation, the lifetime amounts to 2.1969804 μs.

14.4 Transformations of other particles

Unfortunately, the branching ratios of the muon decay are not known precisely. This is different concerning the disintegration of the charged pion for which the experimental relative error $\Delta\varepsilon_{ww}/\varepsilon_{ww}$ is ≈0.005%, and it is therefore worth considering whether there is a similar simple relationship as for the muon.

The charged pion has a mean lifetime of 26.033(5) ns and disintegrates in 99.98770(4)% of the cases into a muon and a muon neutrino. In the Standard Model, the transformation is parameterized by G_F and an additional correcting variable.

With $M = m_\mu - 2m_e$ and $\alpha_{ww} = 1/4\ (1/3 - \alpha)\ \alpha_F$, the deviation $\lvert\varepsilon_{ww}/\varepsilon - 1\rvert$ only amounts to ≈0.002%, if the resonance energy ε is equated to the magnetic energy $5/3\ f_\mu(\alpha_F/\alpha)^{1/2}$. The numbers 1/3 and 5/3, in contrast to the previously discussed muon decay, are experimentally observed Hall fractions. This impressively also confirms that the most accurately known partial lifetime of the charged pion can be related to simple fractions and dimensionless force constants.

Unfortunately, partial decays with relative errors as small as that of the charged pion are rare. However, a high experimental accuracy is mandatory to avoid severe bias in numerical analysis and to achieve good results. Even with a high experimental accuracy, the numerical/experimental agreement of a found resonance channel, through which the process possibly takes place, can still be a numerical coincidence, although the number of possibilities is severely limited by the constricted number of Hall fractions.

The neutron, which is transformed into a proton, an electron and an antielectron neutrino, is surely the most famous representative of nuclear beta decay. Although it is in all likelihood a one-channel decay with a mean lifetime of around 15 min, the experimenters struggle to determine it with high accuracy. The measurement problems are due to the fact that many neutrons, before they disintegrate, are immediately recaptured by the surrounding matter. Depending on the measuring method, the experiments therefore provide contradictory values for the mean lifetime of the neutron outside the nucleus. Recent measurements [14.6] at the Centre for Neutron Research of NIST provided a value of (887.7 ± 2.3) s, with an improved experimental setup. This value is statistically significant different compared to storage experiments and therefore leads to controversial discussions.

Setting $M = 2m_e$ and $\alpha_{ww} = 1/4\ (7/4\alpha_{grav}\alpha)^{1/2}$, then the deviation $|\varepsilon_{ww}/\varepsilon - 1|$ is about 0.04%, if the resonance energy ε is also equated to the magnetic energy $2/5\ f_\mu(\alpha_F\alpha^3)^{1/2}$. The term $(7/4\alpha_{grav}\alpha)^{1/2}$ follows, as explained in Section 11.3, from the lifetime ratio $\tau_{pion\_neutral}/\tau_{neutron}$, and the number 2/5 is a known Hall fraction. The same formalism as for the muon and the pion leads also for the neutron to a good agreement within the experimental error limit of 0.06% without introducing new parameters. Assuming a one-channel decay, the aforementioned process predicts a lifetime of 888.97 s, if CODATA values (2,014) for the natural constants are used.

In Section 11.3, $\alpha^{1/2}$ is deduced as the dimensionless coupling constant for the decay of the neutral pion. In analogy to the disintegration of the neutron, α_{ww} would have to be $1/4\ \alpha^{1/2}$ for the decay of the neutral pion. The lifetime of the neutral pion, which decomposes into two photons in 98.823(34)% of the cases, is $(8.52\pm0.18)\cdot10^{-17}$ s. Since the mean lifetime obviously cannot be determined very accurately, a large experimental error results for $\Delta\varepsilon_{ww}/\varepsilon_{ww}$ of ≈0.59%. Setting $M/m_e = \lambda_{e\_bar}/L_2(\Lambda)$ and $\alpha_{ww} = 1/4\ \alpha^{1/2}$, the deviation $|\varepsilon_{ww}/\varepsilon - 1|$ for the two-photon decay amounts to ≈0.09%, if the resonance energy ε is represented by $3/2\ \lambda_{e\_bar}/L_2(\Lambda)$. The number 3/2 is a Hall fraction observed very rarely.

In the following, further partial lifetimes of the particle zoo are compiled in tabular form according to the same formalism as described earlier. Explanatory text is largely omitted.

Pion(±)

Mean lifetime	$(2.6033\pm0.0005)\cdot10^{-8}$ s
Transformation products	Muon(±) + muon neutrino
Branching ratio	99.98770(4)%
Mass ratio M/m_e	$m_\mu/m_e - 2$
Coupling constant α_{ww}	$1/4\ (1/3 - \alpha)\ \alpha_F$
Resonance energy ε	$5/3\ f_\mu(\alpha_F/\alpha)^{1/2}$
Deviation $\lvert\varepsilon_{ww}/\varepsilon - 1\rvert$	0.002%
Experimental relative error $\Delta\varepsilon_{ww}/\varepsilon_{ww}$	0.005%

Transformation products	Electron(±) + electron neutrino
Branching ratio	0.01230(4)%
Mass ratio M/m_e	2
Coupling constant α_{ww}	$1/4\ 7/5\ \alpha_F\alpha^2$
Resonance energy ε	$5/3\ \alpha\ (4\pi)^{-1}$ [equivalent to $r_{ww} = 3/5\ \lambda_{MI}$]
Deviation $\lvert\varepsilon_{ww}/\varepsilon - 1\rvert$	0.06%
Experimental relative error $\Delta\varepsilon_{ww}/\varepsilon_{ww}$	0.09%

If both processes are combined, an analogous lifetime calculation as in the case for the muon yields 26.0308 ns for then mean lifetime of the charged pion, and 99.98773% or 0.01227% for the two decay probabilities. All calculated values are within the error limits of the PDG data, which are 26.033(5) ns, 99.98770(4)% or 0.01230(4)%, respectively.

The assumption that the transformation of the pion (±) occurs solely via two processes is only approximately fulfilled, since further, but much rarer, transformation channels are experimentally observed. One of the most important channel is the transformation of the charged pion into a neutral pion plus an electron (±) and an electron neutrino (υ_e).

This decay mode of the charged pion is also called the "pion beta decay." It is used to verify the universality of the weak interaction and to clarify a possible violation of the unitarity of the Cabbibo–Kobayashi–Maskawa matrix of the Standard Model. Because of the very small branching ratio of 10^{-8}, the experimental handling of the pion-beta decay is very demanding.

Branching ratio[4]	$1.036(6)\cdot10^{-8}$	
Mass ratio M/m_e	2	
Coupling constant α_{ww}	$1/4\ (9/7\ \alpha_{grav}/\alpha)^{1/2}$	
Resonance energy ε	$f_\mu(\alpha_F\alpha^3)^{1/2}$	
Deviation $\lvert\varepsilon_{ww}/\varepsilon - 1\rvert$	0.05%	[$1.038\cdot10^{-8}$ yields 0.001%]
Experimental relative error $\Delta\varepsilon_{ww}/\varepsilon_{ww}$	0.15%	

The following data can also be evaluated for the very rare decay $\pi^+ \rightarrow e^+\upsilon_e\gamma$

Branching ratio	$7.39(5)\cdot10^{-7}$
Mass ratio M/m_e	2
Coupling constant α_{ww}	$1/4\ (2/3\ \alpha_{grav}/\alpha^2)^{1/2}$
Resonance energy ε	$f_\mu(\alpha_F\alpha^3)^{1/2}$
Deviation $\lvert\varepsilon_{ww}/\varepsilon - 1\rvert$	0.003%
Experimental relative error $\Delta\varepsilon_{ww}/\varepsilon_{ww}$	0.17%

4 The SM prediction [PDG 2006] for the branching ratio is $1.0395(15)\cdot10^{-8}$.

As a comparison, the decay parameters for the neutron are listed again.

Mass ratio M/m_e	2
Coupling constant α_{ww}	$1/4\ (7/4\ \alpha_{grav}\alpha)^{1/2}$
Resonance energy ε	$2/5\ f_\mu(\alpha_F\alpha^3)^{1/2}$

Pion(0)

Mean lifetime	$(8.52\pm0.18)\cdot10^{-17}$ s
Transformation products	Two photons
Branching ratio	98.823(34)%
Mass ratio M/m_e	$\lambda_{e\_bar}/L_2(\Lambda)$
Coupling constant α_{ww}	$1/4\ \alpha^{1/2}$
Resonance energy ε	$3/2\ \lambda_{e\_bar}/L_2(\Lambda)$
Deviation $\lvert\varepsilon_{ww}/\varepsilon - 1\rvert$	0.09%
Experimental relative error $\Delta\varepsilon_{ww}/\varepsilon_{ww}$	0.54%

Kaon(±)

Mean lifetime	$(1.238\pm0.0021)\cdot10^{-8}$ s
Transformation products	Muon(±) + muon neutrino
Branching ratio	63.55(11)%
Mass ratio M/m_e	$m_\mu/m_e + 2$
Coupling constant α_{ww}	$1/4\ (4/5 + \alpha)\ \alpha_F$
Resonance energy ε	$5/2\ f_\mu(\alpha_F/\alpha)^{1/2}$
Deviation $\lvert\varepsilon_{ww}/\varepsilon - 1\rvert$	0.05%
Experimental relative error $\Delta\varepsilon_{ww}/\varepsilon_{ww}$	0.09%

Transformation products	Pion(±) + pion(0)
Branching ratio	20.66(8)%
Mass ratio M/m_e	2
Coupling constant α_{ww}	$1/4\ 4/3\ \alpha_F$
Resonance energy ε	$7/4\ f_\mu(\alpha_F\alpha)^{1/2}$
Deviation $\lvert\varepsilon_{ww}/\varepsilon - 1\rvert$	0.01%
Experimental relative error $\Delta\varepsilon_{ww}/\varepsilon_{ww}$	0.14%

Transformation products	Pion plus + pion plus + pion minus
Branching ratio	5.59(4)%
Mass ratio M/m_e	2
Coupling constant α_{ww}	$1/4\ 4/9\ \alpha_F$

Resonance energy ε	$7/5\ f_\mu(\alpha_F\alpha)^{1/2}$
Deviation $\lvert\varepsilon_{ww}/\varepsilon - 1\rvert$	0.06%
Experimental relative error $\Delta\varepsilon_{ww}/\varepsilon_{ww}$	0.22%

Transformation products	Pion(0) + electron(±) + electron neutrino
Branching ratio	5.07(4)%
Mass ratio M/m_e	$m_\mu/m_e + 2$
Coupling constant α_{ww}	$1/4\ (5/4 + \alpha)\ \alpha_F\alpha$
Resonance energy ε	$1/2\ f_\mu(\alpha_F/\alpha)^{1/2}$
Deviation $\lvert\varepsilon_{ww}/\varepsilon - 1\rvert$	0.16%
Experimental relative error $\Delta\varepsilon_{ww}/\varepsilon_{ww}$	0.24%

Transformation products	Pion(0) + muon(±) + muon neutrino
Branching ratio	3.353(34)%
Mass ratio M/m_e	$m_\mu/m_e - 2$
Coupling constant α_{ww}	$1/4\ (7/5 - \alpha)\ \alpha_F\alpha$
Resonance energy ε	$4/7\ f_\mu(\alpha_F/\alpha)^{1/2}$
Deviation $\lvert\varepsilon_{ww}/\varepsilon - 1\rvert$	0.16%
Experimental relative error $\Delta\varepsilon_{ww}/\varepsilon_{ww}$	0.30%

Transformation products	Pion(±) + pion(0) + pion(0)
Branching ratio	1.761(22)%
Mass ratio M/m_e	$m_\mu/m_e - 2$
Coupling constant α_{ww}	$1/4\ (7/4 - \alpha)\ \alpha_F\alpha$
Resonance energy ε	$3/4\ f_\mu(\alpha_F/\alpha)^{1/2}$
Deviation $\lvert\varepsilon_{ww}/\varepsilon - 1\rvert$	0.04%
Experimental relative error $\Delta\varepsilon_{ww}/\varepsilon_{ww}$	0.35%

Transformation products[5]	Electron(±) + electron neutrino
Branching ratio	$1.582(7)\cdot 10^{-5}$
Mass ratio M/m_e	2
Coupling constant α_{ww}	$1/4\ 2/3\ \alpha_F\alpha^2$
Resonance energy ε	$8/5\ \alpha\ (4\pi)^{-1}$ [equivalent to $r_{ww} = 5/8\ \lambda_{MI}$]
Deviation $\lvert\varepsilon_{ww}/\varepsilon - 1\rvert$	0.39% [$1.56\cdot 10^{-5}$ yields 0.04%]
Experimental relative error $\Delta\varepsilon_{ww}/\varepsilon_{ww}$	0.15%

5 Speculative process analogous to the charged pion. The SM prediction for the branching ratio is $1.57\cdot 10^{-5}$.

$K_L(0)$[6]

Mean lifetime	$(5.116\pm0.021)\cdot10^{-8}$ s
Transformation products	Pion(±) + electron(±) + electron neutrino
Branching ratio	40.55(11)%
Mass ratio M/m_e	2
Coupling constant α_{ww}	$1/4\ 6/5\ \alpha_F$
Resonance energy ε	$2\ f_\mu(\alpha_F\alpha)^{1/2}$
Deviation $\lvert\varepsilon_{ww}/\varepsilon -1\rvert$	0.01%
Experimental relative error $\Delta\varepsilon_{ww}/\varepsilon_{ww}$	0.17%

Transformation products	Pion(±) + muon(±) + muon neutrino
Branching ratio	27.04(7)%
Mass ratio M/m_e	$m_\mu/m_e - 2$
Coupling constant α_{ww}	$1/4\ (3/2 - \alpha)\ \alpha_F$
Resonance energy ε	$1/2\ f_\mu(\alpha_F/\alpha^2)^{1/2}$
Deviation $\lvert\varepsilon_{ww}/\varepsilon -1\rvert$	0.05%
Experimental relative error $\Delta\varepsilon_{ww}/\varepsilon_{ww}$	0.17%

Transformation products	Pion(0) + pion(0) + pion(0)
Branching ratio	19.52(12)%
Mass ratio M/m_e	$m_\mu/m_e + 2$
Coupling constant α_{ww}	$1/4\ (6/5 + \alpha)\ \alpha_F\alpha$
Resonance energy ε	$1/2\ f_\mu(\alpha_F/\alpha)^{1/2}$
Deviation $\lvert\varepsilon_{ww}/\varepsilon -1\rvert$	0.1%
Experimental relative error $\Delta\varepsilon_{ww}/\varepsilon_{ww}$	0.25%

Transformation products	Pion plus + pion minus + pion(0)
Branching ratio	12.54(5)%
Mass ratio M/m_e	2
Coupling constant α_{ww}	$1/4\ 2/3\ \alpha_F\alpha$
Resonance energy ε	$2\ f_\mu(\alpha_F\alpha^2)^{1/2}$
Deviation $\lvert\varepsilon_{ww}/\varepsilon -1\rvert$	0.06%
Experimental relative error $\Delta\varepsilon_{ww}/\varepsilon_{ww}$	0.2%

Transformation products	Pion plus + pion minus
Branching ratio	0.1967(10)%
Mass ratio M/m_e	$m_\mu/m_e - 2$
Coupling constant α_{ww}	$1/4\ (1/3 - \alpha)\ \alpha_F\alpha$

6 $K_L(0)$ is the long-lived or $K_S(0)$ the short-lived version of the neutral kaon.

Resonance energy ε	$4/5\ f_\mu(\alpha_F/\alpha)^{1/2}$
Deviation $\lvert\varepsilon_{ww}/\varepsilon-1\rvert$	0.03%
Experimental relative error $\Delta\varepsilon_{ww}/\varepsilon_{ww}$	0.23%

$K_S(0)$

Mean lifetime	$(0.89564\pm0.00033)\cdot10^{-10}$ s
Transformation products	Pion minus + pion plus
Branching ratio	69.20(5)%
Mass ratio M/m_e	2
Coupling constant α_{ww}	$1/4\ 3/2\ \alpha_F/\alpha$
Resonance energy ε	$2/5\ f_\mu(\alpha_F)^{1/2}$
Deviation $\lvert\varepsilon_{ww}/\varepsilon-1\rvert$	0.03%
Experimental relative error $\Delta\varepsilon_{ww}/\varepsilon_{ww}$	0.03%

Transformation products	Pion(0) + pion(0)
Branching ratio	30.69(5)%
Mass ratio M/m_e	m_μ/m_e-2
Coupling constant α_{ww}	$1/4\ (5/2-\alpha)\ \alpha_F/\alpha$
Resonance energy ε	$3/2\ f_\mu(\alpha_F/\alpha^2)^{1/2}$
Deviation $\lvert\varepsilon_{ww}/\varepsilon-1\rvert$	0.03%
Experimental relative error $\Delta\varepsilon_{ww}/\varepsilon_{ww}$	0.05%

Sigma plus

Mean lifetime	$(0.8018\pm0.0026)\cdot10^{-10}$ s
Transformation products	Proton + pion(0)
Branching ratio	51.57(30)%
Mass ratio M/m_e	2
Coupling constant α_{ww}	$1/4\ 6/5\ \alpha_F/\alpha$
Resonance energy ε	$3/8\ f_\mu(\alpha_F)^{1/2}$
Deviation $\lvert\varepsilon_{ww}/\varepsilon-1\rvert$	0.09%
Experimental relative error $\Delta\varepsilon_{ww}/\varepsilon_{ww}$	0.23%

Transformation products	Neutron + pion plus
Branching ratio	48.31(30)%
Mass ratio M/m_e	m_μ/m_e-2
Coupling constant α_{ww}	$1/4\ (3/8-\alpha)\ \alpha_F/\alpha$
Resonance energy ε	$1/2\ f_\mu(\alpha_F/\alpha^2)^{1/2}$
Deviation $\lvert\varepsilon_{ww}/\varepsilon-1\rvert$	0.03%
Experimental relative error $\Delta\varepsilon_{ww}/\varepsilon_{ww}$	0.24%

Sigma minus

Mean lifetime	$(1.479\pm0.011)\cdot10^{-10}$ s
Transformation products	Neutron + pion minus
Branching ratio	99.848(5)%
Mass ratio M/m_e	2
Coupling constant α_{ww}	1/4 7/5 α_F/α
Resonance energy ε	2/5 $f_\mu(\alpha_F)^{1/2}$
Deviation $\lvert\varepsilon_{ww}/\varepsilon-1\rvert$	0.04%
Experimental relative error $\Delta\varepsilon_{ww}/\varepsilon_{ww}$	0.19%

Sigma (0)

The neutral sigma particle has a mean lifetime of $(7.4\pm0.7)\cdot10^{-20}$ s and transforms almost completely into a neutral lambda particle plus a photon. For consistency reasons to the neutral pion decay, the dimensional coupling constant α_{ww} must be proportional to 1/4 2/3 α^{-1}. This leads with $M/m_e = \lambda_{e\_bar}/L_2(\Lambda)$ to an approximate resonance with the energy $\varepsilon = \lambda_{e\_bar}/L_2(L)$. The deviation is 2.1% (!), which lies within the experimental error band of 2.4%. Nevertheless, this process is very speculative due to the large experimental error involved.

Xi minus

Mean lifetime	$(1.639\pm0.015)\cdot10^{-10}$ s
Transformation products	Lambda + pion minus
Branching ratio	99.887(35)%
Mass ratio M/m_e	2
Coupling constant α_{ww}	1/4 1/3 α_F/α
Resonance energy ε	1/5 $f_\mu(\alpha_F)^{1/2}$
Deviation $\lvert\varepsilon_{ww}/\varepsilon-1\rvert$	0.07%
Experimental relative error $\Delta\varepsilon_{ww}/\varepsilon_{ww}$	0.24%

Xi(0)

Mean lifetime	$(2.90\pm0.09)\cdot10^{-10}$ s
Transformation products	Lambda + pion(0)
Branching ratio	99.524(12)%
Mass ratio M/m_e	2
Coupling constant α_{ww}	1/4 α_F/α
Resonance energy ε	2/5 $f_\mu(\alpha_F)^{1/2}$
Deviation $\lvert\varepsilon_{ww}/\varepsilon-1\rvert$	0.05%
Experimental relative error $\Delta\varepsilon_{ww}/\varepsilon_{ww}$	0.8%

Lambda

Mean lifetime[7]	$(2.632 \pm 0.020) \cdot 10^{-10}$ s
Transformation products	Proton + pion minus
Branching ratio	63.9(5)%
Mass ratio M/m_e	2
Coupling constant α_{ww}	$1/4\ 3/7\ \alpha_F/\alpha$
Resonance energy ε	$2/7\ f_\mu(\alpha_F)^{1/2}$
Deviation $\lvert\varepsilon_{ww}/\varepsilon - 1\rvert$	0.02%
Experimental relative error $\Delta\varepsilon_{ww}/\varepsilon_{ww}$	0.4%

Transformation products	Neutron + pion(0)
Branching ratio	35.8(5)%
Mass ratio M/m_e	2
Coupling constant α_{ww}	$1/4\ 7/4\ \alpha_F/\alpha$
Resonance energy ε	$2/3\ f_\mu(\alpha_F)^{1/2}$
Deviation $\lvert\varepsilon_{ww}/\varepsilon - 1\rvert$	0.08%
Experimental relative error $\Delta\varepsilon_{ww}/\varepsilon_{ww}$	0.5%

Since other decays are very rare, a mean lifetime of $2.638055 \cdot 10^{-10}$ s can be calculated by means of the two processes given earlier. The calculated decay probabilities are $(4/5)^2 = 64\%$ and $(3/5)^2 = 36\%$, respectively.

Neutral D meson

Mean lifetime	$(410.1 \pm 1.5) \cdot 10^{-15}$ s
Transformation products	Kaon minus + pion plus
Branching ratio	3.93 (4)%
Mass ratio M/m_e	2
Coupling constant α_{ww}	$1/4\ 2/3\ \alpha_F/\alpha^2$
Resonance energy ε	$5/3\ f_\mu(\alpha_F)^{1/2}$
Deviation $\lvert\varepsilon_{ww}/\varepsilon - 1\rvert$	0.07%
Experimental relative error $\Delta\varepsilon_{ww}/\varepsilon_{ww}$	0.34%

Transformation products	Kaon minus + Kaon plus
Branching ratio	0.401(7)%
Mass ratio M/m_e	2

7 Because of the unexpectedly high mean lifetime, the Standard Model introduced the quantum number "strangeness." The lambda particle is the lightest strange baryon. It cannot decay strongly because of a combination of baryon number conservation and strangeness.

Coupling constant α_{ww}	$1/4\ 1/4\ \alpha_F/\alpha$
Resonance energy ε	$9/5\ f_\mu(\alpha_F\alpha)^{1/2}$
Deviation $\lvert\varepsilon_{ww}/\varepsilon - 1\rvert$	0.2%
Experimental relative error $\Delta\varepsilon_{ww}/\varepsilon_{ww}$	0.5%

Transformation products	Pion plus + pion minus
Branching ratio	0.1421(25)%
Mass ratio M/m_e	2
Coupling constant α_{ww}	$1/4\ 2/7\ \alpha_F/\alpha$
Resonance energy ε	$5/2\ f_\mu(\alpha_F\alpha)^{1/2}$
Deviation $\lvert\varepsilon_{ww}/\varepsilon - 1\rvert$	0.01%
Experimental relative error $\Delta\varepsilon_{ww}/\varepsilon_{ww}$	0.5%

Transformation products	Kaon plus + pion minus
Branching ratio	0.01399(27)%
Mass ratio M/m_e	2
Coupling constant α_{ww}	$1/4\ 7/2\ \alpha_F$
Resonance energy ε	$4/3\ f_\mu(\alpha_F\alpha)^{1/2}$
Deviation $\lvert\varepsilon_{ww}/\varepsilon - 1\rvert$	0.002%
Experimental relative error $\Delta\varepsilon_{ww}/\varepsilon_{ww}$	0.6%

The investigation of decays of neutral D mesons in pairs of kaons or pions is an active field of research because of CP symmetry violation. Therefore, data will probably soon be available with higher accuracy for previously mentioned processes.

Tauon

Mean lifetime	$(290.3\pm0.5)\cdot10^{-15}$ s
Transformation products	Pion minus + pion(0) + tau neutrino
Branching ratio	25.52(9)%
Mass ratio M/me	2
Coupling constant α_{ww}	$1/4\ 4\ \alpha_F/\alpha^2$
Resonance energy ε	$1/5\ f_\mu(\alpha_F/\alpha)^{1/2}$
Deviation $\lvert\varepsilon_{ww}/\varepsilon - 1\rvert$	0.03%
Experimental relative error $\Delta\varepsilon_{ww}/\varepsilon_{ww}$	0.13%

Transformation products	e^- + antielectron neutrino + tau neutrino
Branching ratio	17.83(4)%
Mass ratio M/m_e	$m_\mu/m_e - 2$
Coupling constant α_{ww}	$1/4\ (3/5 - \alpha)\ \alpha_F/\alpha^2$
Resonance energy ε	$1/5\ f_\mu(\alpha_F/\alpha^3)^{1/2}$

Deviation $\lvert\varepsilon_{ww}/\varepsilon - 1\rvert$	0.08%	
Experimental relative error $\Delta\varepsilon_{ww}/\varepsilon_{ww}$	0.1%	

Rho meson resonance

Mean lifetime	$4.45(3)\cdot 10^{-24}$ s	[147.8 ± 0.9 MeV]
Transformation products	Two pions	
Branching ratio	100%	
Mass ratio M/m_e	$\lambda_{e\_bar}/L_2(\Lambda)$	
Coupling constant α_{ww}	$1/4\ 2\ \alpha^{-2}$	
Resonance energy ε	$7/4\ \lambda_{e\_bar}/L_2(L)$	
Deviation $\lvert\varepsilon_{ww}/\varepsilon - 1\rvert$	0.01%	
Experimental relative error $\Delta\varepsilon_{ww}/\varepsilon_{ww}$	0.15%	

Transformation products	Electron + positron	
Branching ratio	$4.72(5)\cdot 10^{-5}$	
Mass ratio M/m_e	$\lambda_{e\_bar}/L_2(\Lambda)$	
Coupling constant α_{ww}	$1/4\ 8/9\ \alpha^{-1}$	
Resonance energy ε	$6/5\ \lambda_{e\_bar}/L_2(L)$	
Deviation $\lvert\varepsilon_{ww}/\varepsilon - 1\rvert$	0.1%	
Experimental relative error $\Delta\varepsilon_{ww}/\varepsilon_{ww}$	0.4%	

Omega meson resonance

Mean lifetime	$7.75(7)\cdot 10^{-23}$ s	[8.49 ± 0.08 MeV]
Transformation products	Three pions	
Branching ratio	89.2(7)%	
Mass ratio M/m_e	$\lambda_{e\_bar}/L_2(\Lambda)$	
Coupling constant α_{ww}	$1/4\ 4/3\ \alpha^{-2}$	
Resonance energy ε	$3\ \lambda_{e\_bar}/L_2(L)$	
Deviation $\lvert\varepsilon_{ww}/\varepsilon - 1\rvert$	0.1%	
Experimental relative error $\Delta\varepsilon_{ww}/\varepsilon_{ww}$	0.4%	

Transformation products	Electron + positron	
Branching ratio	$7.28(14)\cdot 10^{-5}$	
Mass ratio M/m_e	$\lambda_{e\_bar}/L_2(\Lambda)$	
Coupling constant α_{ww}	$1/4\ 5/9\ \alpha^{-1}$	
Resonance energy ε	$7/4\ \lambda_{e\_bar}/L_2(L)$	
Deviation $\lvert\varepsilon_{ww}/\varepsilon - 1\rvert$	0.5%	
Experimental relative error $\Delta\varepsilon_{ww}/\varepsilon_{ww}$	0.7%	

Phi meson resonance

Mean lifetime	$1.54(1)\cdot10^{-22}$ s [4.266 ± 0.031 MeV]
Transformation products	Kaon plus + kaon minus
Branching ratio	48.9(5)%
Mass ratio M/m_e	$\lambda_{e\_bar}/L_2(\Lambda)$
Coupling constant α_{ww}	$1/4\ 1/4\ \alpha^{-2}$
Resonance energy ε	$9/5\ \lambda_{e\_bar}/L_2(L)$
Deviation $\lvert\varepsilon_{ww}/\varepsilon -1\rvert$	0.3%
Experimental relative error $\Delta\varepsilon_{ww}/\varepsilon_{ww}$	0.4%

Transformation products	Electron + positron
Branching ratio	$2.954(30)\cdot10^{-4}$
Mass ratio M/m_e	$\lambda_{e\_bar}/L_2(\Lambda)$
Coupling constant α_{ww}	$1/4\ 3/8\ \alpha^{-1}$
Resonance energy ε	$6/5\ \lambda_{e\_bar}/L_2(L)$
Deviation $\lvert\varepsilon_{ww}/\varepsilon -1\rvert$	0.3%
Experimental relative error $\Delta\varepsilon_{ww}/\varepsilon_{ww}$	0.4%

The preceding examples demonstrate that, independent of any unit system, many partial decays within their experimental uncertainties can be reduced to a few simple Hall fractions and dimensionless constants. The question whether the observed universality with the good experimental agreements is a coincidence is left to the reader.

In any case, some complicated phenomena of particle physics can be quantitatively cataloged, and interpreted by numerical testing using a simple, universal model without introducing extra free parameters apart from a number of Hall fractions. Whether the observed pattern can be used to draw valid and fruitful conclusions for the physical understanding of the phenomena of particle physics will only become apparent when more precise, independent experiments with improved statistics are available. Any ambiguities in the choice of Hall fractions can then be further limited.

14.5 The exotic positronium atom

The positronium (Ps) is a bound state between an electron and its antiparticle, the positron, before the electron and the positron annihilate to two photons after about 10^{-10} s.

The previous Section 14.4 has shown that particle transformations can phenomenologically be cataloged by means of the old Fermi theory and Hall fractions. Is this principle so universal that by analogy the mean lifetime of Ps can be characterized in the same way? This Section is intended to investigate this question.

Relation 14.4 can be transformed into

Relation 14.5 $\tau_{tot} = (3/4)\, \varepsilon_{ww}{}^4\, \alpha_{ww}{}^{-2}\, h/(m_e c^2)$

when $M/m_e = 2$ and $B_i = 1$ is used. The reciprocal energy $\varepsilon_{ww}{}^{-1}$ is equivalent to the interaction length r_{ww} in units of λ_{e_bar}. The length r_{ww} shall be chosen so that it corresponds to the relevant Bohr radius of the Ps, that is, twice the Bohr radius $2a_H$ of the electron in the hydrogen atom. Since $a_H = \lambda_{e\_bar}/\alpha$, it follows for $\varepsilon_{ww} = \alpha/2$.

By numerical testing, the coupling constant $\alpha_{ww} = (1+3/8)\, \alpha_F$ can be found, which approximates the measured lifetime of the para-Ps very well. If the dimensionless time $x_1 = (3/4)\, \varepsilon_{ww}{}^4\, \alpha_{ww}{}^{-2}$, which can be calculated solely by means of geometrized values of α and α_F, is multiplied by the reference time $h/(m_e c^2)_{CODATA}$, a mean lifetime of 125.09 ps or a decay rate of 7.9940 ns^{-1} is obtained.

Experimentally, a value of 7.9909(17) ns^{-1} was measured by A. H. Al-Ramadhan and D. W. Gidley [14.8]. Interestingly, the fraction 3/8 is also present in the phenomenological description of the decay processes of the sigma plus particle, whose lifetime (≈80 ps) is of the same order of magnitude as that of para-Ps. Also in the fundamental Relation 5.5, the numbers 3 and 8 arise, possibly indicating a symmetry property.[8]

Table 14.1: Decay rates of Ps.

Particle	r_{ww}/a_H	α_{ww}	Calculated	Experiment	BS-QED
Para-Ps	2	$(1+3/8)\,\alpha_F$	7.9940 ns^{-1}	7.9909(17) ns^{-1} [14.8]	7.9896 ns^{-1}
Ps^-	3/2	$(5/4)\,\alpha_F$	2.0904 ns^{-1}	2.089(15) ns^{-1} [14.9]	2.0871 ns^{-1}
Ortho-Ps	3	$(5/2)\,\alpha_F\alpha$	7.0987 μs^{-1}	7.0482(16) μs^{-1} [14.10]	7.0382 μs^{-1}

Note: The decay rates of para-Ps, ortho-Ps and Ps^- are listed as a function of α_{ww} and ε_{ww} according to Relation 14.5. As a comparison, the experimental and the theoretical values calculated according to BS-QED (bound state QED) are listed. The errors of the measured values are indicated in round brackets.

Similarly, the decay rates of ortho-Ps and the negatively charged Ps ion (Ps^-) were determined by numerical trial and error. Table 14.1 summarizes the results. It would be helpful if there were additional, independent measurements available, so that mean values can be calculated similar to the PDG data. It also needs to be clarified why the measured decay rate of ortho-Ps approximated the calculated QED value over the last 40 years in an even better manner. Measuring the decay rate of ortho-Ps is very difficult because extrapolations and corrections are necessary to get the vacuum value.

8 In the Standard Model of particle physics there are three gauge bosons (W^+, W^-, Z^0) for the weak interaction and eight gauge bosons (gluons) for the strong interaction.

14.5.1 Hyperfine splitting

According to the latest measurements, the hyperfine splitting $(\Delta f)_{Ps}$ of Ps is 203.3942(21) GHz [14.7]. Using

Ansatz 14.5 $(\Delta f)_{Ps} = (7/12)\,\alpha^4\,\{1 + \alpha/\pi - (\alpha_F/\pi)^{1/2}\}\,(m_e c^2/h)$

a singlet-triplet splitting of 203.3949 GHz can be calculated when geometrized values for α and α_F and CODATA values for m_e, c and h are used. The physical meaning of $(\alpha_F/\pi)^{1/2}$ is not clear. It is striking that a similar term also occurs in Ansatz 11.3.

14.6 Magnetic moments of heavy particles

Not only the calculation of particle masses, but also the calculation of the magnetic moments of baryons poses difficult problems for physicists, and is waiting for a unified explanation. For heavy particles, the Dirac equation, which predicts g = 2 for the spin g-factor, does not apply due to experimental findings. According to QED, the uncharged neutron should have no magnetic moment at all. But to the surprise of all physicists, a magnetic moment was measured. In the electrically neutral neutron, there are apparently charges that shield each other and generate a magnetic moment despite charge neutrality.

The current doctrine supports the view that magnetic moments are caused by quarks confined in the nucleus with fractional charges and their interactions. But quantum chromodynamics – a theory that contains six quarks, eight gluons and the interactions between them – cannot give a clear picture of magnetic moments of heavy particles. Even using the experimental data of the magnetic moments of proton, neutron and lambda, the magnetic moments of the remaining baryons can only be calculated with an accuracy of ±20%. Lattice calculations using high-performance parallel computers do not lead to the desired goal either. In short, modern physics theorists are unable to theoretically derive nuclear magnetic moments and went astray in this respect.

As an additional test, this section will clarify whether magnetic moments can simply be calculated in terms of fundamental energy quanta and Hall fractions without a supercomputer.

The relativistic quantum theory defines the mass M, the spin S, the charge Q and the spin g-factor as the relevant factors for the magnetic moment μ. These parameters are related by the generally valid

Formula 14.3 $\mu_{particle} = g\,S\,Q\,h_{bar}/(2M)$

For example, the magnetic moment of the electron $\mu_e \approx -\mu_{Bohr}$ results from Formula 14.3, with $g \approx -2$, $Q = q_e$ and $S = 1/2$.

If g is set to 1, that is, when heavy particles are classically considered to be rotating, uniformly charged spheres, with $Q = Z\ q_e$ and $S = 1/2$, Formula 14.3 yields by a simple transformation

Relation 14.6 $\mu_{particle}/\mu_{Bohr} = Z\ (M/m_e)^{-1}\ (1/2)$

Interestingly, when fractional values or Hall fractions are allowed for Z, magnetic moments of heavy particles in the fundamental unit of μ_{Bohr} can be calculated. For the mass M not the particle mass itself may be used, but, depending on the particle, a sum of the fundamental mass building blocks $L_2(L)^{-1}h_{bar}c^{-1}$ and m_e has to be inserted. Table 14.2 summarizes such calculations that are independent of any unit system. For $\lambda_{e\_bar}/L_2(L)$, of course, the geometrized value $2^{16}\ \pi^{-5}$ must be used. All results were obtained by numerical trial and error.

Table 14.2: Magnetic moments.

Particle	Charge Z	M/m_e	$\mu_{particle}/\mu_{Bohr}/10^{-3}$	
			Relation 14.6	Experiment
Proton	+2/3	$\lambda_{e\_bar}/L_2(L) + 5$	+1.520985	+1.5210322053(46)
Neutron	−4/9	$\lambda_{e\_bar}/L_2(L) - 1$	−1.042532	−1.04187563(25)
Deuteron	+1/5	$\lambda_{e\_bar}/L_2(L)$	+0.466949	+0.4669754554(26)
Lambda	−1/7	$\lambda_{e\_bar}/L_2(L)$	−0.3335	−0.3339(22)
Sigma plus	+4/7	$\lambda_{e\_bar}/L_2(L)$	+1.334	+1.339(5)
Sigma minus	−2/7	$\lambda_{e\_bar}/L_2(L) + 9$	−0.640	−0.632(13)
Xi(0)	−2/7	$\lambda_{e\_bar}/L_2(L) - 5$	−0.683	−0.681(8)
Xi minus	−1/7	$\lambda_{e\_bar}/L_2(L) - 9$	−0.3482	−0.3544(14)
Omega minus	−1/2	$\lambda_{e\_bar}/L_2(L) + 9$	−1.120	−1.100(27)

Note: The table lists magnetic moments of some particles in units of μ_{Bohr} and the comparison of the predictions of Relation 14.6 with experimental values. The experimental values of the proton, the neutron and the deuteron magnetic moment are CODATA values (2014). The rest of the data are extracted from the listing of the Particle Data Group [14.1], which tabulates magnetic moments in units of μ_N. The conversion to the unit μ_{Bohr} is achieved by dividing the PDG value by m_p/m_e. Contrary to convention, the sign of the magnetic moment is not assigned to the g-factor, but to Z

It seems that the two energies $L_2(L)^{-1}h_{bar}c$ and m_ec^2 merge into quasiparticles, which become noticeable in the measurement of the magnetic moment with a fractional charge. The energy $L_2(L)^{-1}h_{bar}c$, unlike the rest energy of the electron, does not appear as a free particle and cannot be extracted from the nuclei. It can only exist within compound of seven electrons as a muon in the real world for a short time period.

Although many electrons are formed in particle reactions, today's teaching is that there exist no electrons in the nuclei.

The concept of quasiparticles (bound states) with fractional charges and masses composed of the energy quantum $L_2(L)^{-1}h_{bar}c$ and an integral multiple of the rest energy m_ec^2 is quite successful in the theoretical calculation of nuclear magnetic moments. The deviation from the measured value $|\mu_{exp}/\mu_{theo} - 1|$ is 0.003% for the proton, 0.006% for the deuteron and 0.06% for the neutron. For the remaining particles, except for the xi minus, the predictions are essentially within the experimental error limits.

In quantum theory, symmetries usually generate simple formulas with natural or fractional numbers. Is it conceivable that some kind of symmetry is also responsible for the results of Table 14.2? Why is the deviation from the measured value to the calculation value for the neutron ten times greater than for the proton? Could the larger deviation for the neutron be related to the fact that the nuclear magnetic moment of the neutron is reduced by the external magnetic field applied during the measurement?

14.7 Charge radii of light hadrons

In this section we try to interpret, in analogy to Relation 14.6, the mean square charge radii ($<r^2_{hadron}>^{1/2}$) of the lightest hadrons by a sum of integral multiples of the energies $L_2(\Lambda)^{-1}h_{bar}c$ and m_ec^2.

An energy density can be converted into the equivalent form $\varepsilon^4 (h_{bar}c)^{-3}$, if the unit volume is understood as an inverse energy to the power of three. This corresponds to the usual method in particle physics, to express physical units as powers of energy. Using this principle, Formula 9.2 can be modified to

Relation 14.7 $R = (\alpha/8\pi)^{1/4} \varepsilon^{-1} (h_{bar}c)$

which links the radius R of a sphere with a homogeneously charged shell to an energy that is equivalent to an energy density in the sense discussed earlier. Setting $\varepsilon = L_2(\Lambda)^{-1}h_{bar}c + m_ec^2$, Relation 14.7 yields after some algebraic manipulations

Relation 14.8 $R/\lambda_{e\_bar} = (\alpha/8\pi)^{1/4} (\lambda_{e\_bar}/L_2(\Lambda) + 1)^{-1}$

For integral multiples of $L_2(\Lambda)^{-1}h_{bar}c$ or m_ec^2, Relation 14.8 must be modified accordingly. The ratios R/λ_{e_bar} and $\lambda_{e\_bar}/L_2(\Lambda)$ are number constants and independent of natural constants, if $\alpha = \alpha_{geom}$ and $\lambda_{e\_bar}/L_2(\Lambda) = 2^{14} \pi^{-17/3}$ are applied. The connection of Relation 14.8 to the physical reality is achieved by multiplying the dimensionless number R/λ_{e_bar} with the reduced Compton length $\lambda_{e\_bar\_Codata}$.

The question is, whether experimental mean square charge radii of nuclear charge distributions can be related to an energy ε, as it has been said. Table 14.3 lists the results for some hadrons, which were found by numerical trial and error. All measured charge radii can be explained surprisingly well with simple combinations of integral multiples of the energy quanta $L_2(\Lambda)^{-1}h_{bar}c$ and m_ec^2. It is astonishing to see the excellent experimental agreement for the proton and the charged pion. Unfortunately, no data are available for other particles such as sigma plus, xi minus or omega minus. Only more precise data and additional experiments can clarify whether the interpretation of charge radii, using integral multiples of $L_2(\Lambda)^{-1}h_{bar}c$ and m_ec^2 according to Relation 14.8, is compatible with experimental findings, and Relation 14.8 can be investigated for how charge building blocks are arranged.

Relation 14.8 is rendering the idea of Section 8.1 obsolete that the charge radius of the proton might be related to the scattering length $L_2(L_p)$. Which interpretation might be right? According to recent investigations [14.12], that is, using just proton scattering data without model-dependent assumptions, the proton's mean charge radius is closer to ≈0.87 fm than it is to ≈0.90 fm, which favors Relation 14.8 that can additionally be used to calculate charge radii of other hadrons as well.

Table 14.3: Charge radii.

Particle	**$\varepsilon/(m_ec^2)$**	**$R/\lambda_{e\_bar}/10^{-3}$**	**R (fm)** Relation 14.8	**$<r^2_{hadron}>^{1/2}$ (fm)** Experiment
Charged kaon	$3\ \lambda_{e\_bar}/L_2(\Lambda) + 9$	1.555553	0.601	0.580(40)
Charged pion	$3\ \lambda_{e\_bar}/L_2(\Lambda)$	1.742522	0.673	0.672(8)
Sigma minus	$3\ \lambda_{e\_bar}/L_2(\Lambda) - 9$	1.980575	0.765	0.780(100)
Proton	$2\ \lambda_{e\_bar}/L_2(\Lambda) + 8$	2.252757	0.870	0.879(11) [14.11] 0.871(10) [14.12]
Deuteron	$1\ \lambda_{e\_bar}/L_2(\Lambda) - 1$	5.445747	2.103	2.130(10) [14.13]

Note: The table lists charge radii of five hadrons in units of λ_{e_bar} and the comparison of the predictions of Relation 14.8 to experimental data. The experimental values $<r^2_{hadron}>^{1/2}$ without references are retrieved from the listing of the Particle Data Group [14.1]. The CODATA value (2014) of $<r^2_{proton}>^{1/2}$ is 0.8751(61) fm

References

[14.1] Particle data group. http://pdg.lbl.gov, queried on July 15, 2015

[14.2] H. L. Stormer, D. C. Tsui and A. C. Gosshard. The fractional quantum Hall effect. Review of Modern Physics 71 (1999) S298–S305

[14.3] Observation of J/ψp resonances consistent with pentaquark states. LHCb collaboration. arXiv:1507.03414v2 [hep-ex]

[14.4] M. Fael, L. Mercolli and M. Passera. Radiative μ and τ leptonic decays at NLO. arXiv:1506.03416 [hep-ph]

[14.5] D. J. Griffiths. Introduction to elementary particles. Wiley, N.Y., 1987
[14.6] A. T. Yue, M. S. Dewey and D. M. Gilliam. Improved determination of the neutron lifetime. Physical Review Letters 111 (2013) 222501
[14.7] A. Ishida,T. Namba, S. Asai et al. New precision measurement of hyperfine splitting of positronium. Physics Letters B 734 (2014) 338–344
[14.8] A. H. Al-Ramadhan and D. W. Gidley. New precision measurement of the decay rate of singlet positronium. Physical Review Letters 72 (1994) 1632–1635
[14.9] Frank Fleischer, K. Degreif, G. Gwinner et al. Measurement of the decay rate of the negative ion of positronium. Physical Review Letters 96 (2006) 063401 [4 pages]
[14.10] J. S. Nico, D. W. Gidley, A. Rich and P. W. Zitzewitz. Precision measurement of the orthopositronium decay rate using the vacuum technique. Physical Review Letters 65 (1990) 1344–1347
[14.11] J. Arrington and Ingo Sick. Evaluation of the proton charge radius from electron proton scattering. J. Phys. Chem. Ref. Data 44 (2015) 031204 [6 pages]. arXiv:1505.02680 [nucl-ex]
[14.12] Richard J. Hill and Gil Paz. Model independent extraction of the proton charge radius from electron scattering. Physical Review D 82 (2010) 113005 [18 pages]
[14.13] I. Sick and D. Trautmann. On the rms-radius of the deuteron. Physics Letters B 375 (1996) 16–20

15 Interpretation of hydrogen-like systems with α as a number constant

> [There is] some mathematical quality in Nature, a quality which the casual observer of Nature would not suspect, but which nevertheless plays an important role in Nature's scheme.
> (*Paul A. M. Dirac*)

Hydrogen-like systems such as ^1H, ^2H or $^3He^+$, which contain only one single electron, play a central role in precision atomic physics. In the physics community, these systems solely bound by photons are viewed as completely solved. It may therefore be surprising to devote a separate chapter to such systems and to reopen the beginnings of quantum mechanics, although the interpretation of spectroscopic data by existing models is excellent.

Whoever changes the fine-structure constant α as significantly as the redefinition of α_{geom} according to Definition 5.3 has the duty to check the implementation of the geometrized fine-structure constant on the simplest bound systems. As is well known, this constant is compulsorily linked with bound systems because it defines the energy levels and their splittings.

This chapter also aims to establish a link to particle physics, which in its present form does not provide any clues as to why atoms exist. Although the question of existence is not answered in what follows in this chapter, but at least a possible link is demonstrated, as the reevaluation of spectroscopic data is based on dimensionless quantities of previous chapters dealing with particle physics. A connection between atomic and particle physics is obvious because hyperfine splittings (hfs) showing in atomic spectra are caused by nuclei.

Today's theory of hydrogen-like systems is based on a complicated, singular mathematics which enables to absorb infinities at high energies into quantities to be measured. Physically, this procedure cannot be justified. On the one hand, the theory implies that the charge q_e and the electron mass m_e must be assumed to be infinite before a measurement, that is, without a radiation field, and consequently the electron does not exist. On the other hand, there is a great danger of a vicious circle, that is, that what is measured is also confirmed. The more accurately the quantities q_e and m_e are tabulated, the more they look like a circular conclusion and the theory determines what is measurable.

Today, the fine-structure constant at low energies is de facto determined from a divergent quantum electrodynamics (QED) approach for the anomalous magnetic moment of the electron a_e, and any deviation of α greater than 10^{-10} is considered noncompetitive. This eliminates "bad" data, creates consistency and finds a solution to the problem that QED does not predict numbers that can be compared to the experiment, but provides only formulas in which α can be substituted. Values for α, which result from atomic spectra, are therefore no longer used by the CODATA

https://doi.org/10.1515/9783110612387-015

group today. The hunt for more and more decimal places increasingly takes on bizarre features. The influence of the theory is complete if, as a consequence of quantum physics, the constant α is additionally related to the energy scale of the measurement.

In this chapter, by means of hypotheses of previous chapters experimental data are consistently ascribed to number constants, which are related to physical facts. In addition, all considerations are based on the principle of simplicity. The connection of the dimensionless expressions to physical reality is achieved by compiled CODATA values. This is necessary because concepts without units are mathematically very appealing, but without standardized practical units, that is, a classical language, they have no physical meaning.

15.1 The correction or scaling factor γ

The interpretation of the spectral lines of hydrogen-like systems is based on the Rydberg constant R_∞, which by definition is $\alpha^2/2\ \{m_ec/h\}$ for infinitely heavy nuclei. Thus, it is a product of the dimensionless fundamental constant α squared and the "practical constant" $\{m_ec/h\}$ with the arbitrary macroscopic unit m^{-1} or cm^{-1}, respectively. Other units for R_∞ such as J, Hz and eV can be formed by multiplying $\alpha^2/2$ by other "practical constants" (conversion factors) such as $\{m_ec^2\}$, $\{m_ec^2/h\}$ or $\{m_ec^2/q_e\}$. In fact, in most cases the quantity cR_∞ is measured with the unit Hz. From the definition of R_∞ can be ascertained that every change in α leads to a change in R_∞, and thus to different positions of the spectral lines because the Rydberg constant is the scaling factor of atomic levels. However, a measurement must not have a preferential status with regard to the system to be measured.

If α is replaced by α_{geom}, this implies to a first approximation that a correction factor γ must be introduced so that the wavelengths of the spectral lines remain unchanged. In the process of geometrization of the electron mass m_e in Section 6.1, c and h are invariants, and it is therefore obvious for reasons of consistency to regard c and h as invariant here as well, and to assign the correction factor γ exclusively to the electron mass m_e. This assignment also pays heed to the fact that a moving mass is affected differently from a moving electric charge which has the same value for all observers. If the value α_{geom} is used instead of α_{Codata}, the correction factor γ is equal to the quantity $(\alpha_{Codata}/\alpha_{geom})^2$ or ≈ 1.0035918, respectively. This "experimental" value is independent of how α_{Codata} was derived, since it follows from

Equation 15.1 $$2R_\infty = \alpha_{Codata}{}^2\,(m_ec/h)_{Codata} = \gamma\,\alpha_{geom}{}^2\,(m_ec/h)_{Codata}$$

The introduction of γ cancels the special position of the mass m_e in the Rydberg constant. The dimensionless product $\gamma\,\alpha_{geom}{}^2$ must necessarily be traceable to number constants because $(m_ec/h)_{Codata}$ merely represents the physical reference unit (measuring unit) without a special status.

The empirical Rydberg formula $\Delta f = cR_\infty (1/n^2 - 1/m^2)$ determines the frequencies or the wavelengths of the spectral lines and is also the origin of the Rydberg constant. Like many properties of the elements, the empirical Rydberg formula depends on natural numbers whose existence quantum mechanics cannot justify. Relative intensities of emission lines of atomic hydrogen can also be characterized by natural numbers. For more details about this subject see Table 3.3. Might it be possible that γ is also ascribable to natural numbers? Therefore, on an ad hoc basis

Definition 15.1 $\gamma^{-2} = 1 - 140^{-1}$

shall apply. From Definition 15.1 a value of ≈1.0035907 for γ can be calculated, which differs only ≈1 ppm from the "experimental" value $(\alpha_{Codata}/\alpha_{geom})^2$. If $(v/c)^2$ is equated to 140^{-1} or $(4{\cdot}5{\cdot}7)^{-1}$, then Definition 15.1 can be associated with a relativistic, Lorentz-invariant approach. The Lorentz invariance is a fundamental principle of the theory of relativity with the requirement that the speed of light in vacuum is always constant.

15.2 Phenomenological extension of the Dirac–Sommerfeld model

In 1916, Arnold Sommerfeld, as a generalization of the Bohr model, developed a formula for the electron describing elliptical Kepler orbits around the atomic nucleus. His theory is consistent with the special theory of relativity and surprisingly agrees with the Dirac theory for a single point-like electron, which was developed later, and which can also be understood only relativistically. In both approaches, however, the quantum numbers have a different meaning.

The binding energy E_D of an electron in the field of an infinitely heavy nucleus with the principal quantum number n and the total angular momentum j is given in both theories by [CODATA-2014, Formulas 22, 23 and 24]

Formula 15.1 $E_D/(m_ec^2) = f(n, j) - 1$

with

$$f(n, j)^{-2} = 1 + \alpha^2 / (n-\delta)^2$$

and

$$\delta = k - (k^2 - \alpha^2)^{1/2} \text{ with } k = j+1/2$$

In bound state, the spin s of the electron itself is not observable, since spin and orbital angular momentum l couple to the total angular momentum $j = s + l$. Formula 15.1 describes the movement of a relativistic electron in the field of an infinitely heavy nucleus without a nuclear spin.

In fact, hydrogen-like systems are two-body problem, where the nucleus (m_1) and the electron (m_2) move relative to the common center of mass of the system. Unfortunately, to this day there is no reliable treatment of the relativistic two-body problem available that fully takes into account the relativistic dynamics and the interactions between both components. Approximations are needed. Even for the simplest atom such as hydrogen there is no exact quantitative description on hand. The most obvious approximation is to refer the coordinates to a common origin, so that an "effective" problem for one body arises with the reduced mass m_R given by

Relation 15.1 $1/m_R = 1/(m_1\gamma_1) + 1/(m_2\gamma_2)$

Setting $m_1 = M$, $\gamma_1 = 1$, $m_2 = m_e$ and $\gamma_2 = \gamma$,

Relation 15.2 $m_R = \gamma\, m_e/(1 + \gamma\, m_e/M)$

can be deduced. Using Formula 15.1 and Relation 15.2 yields

Relation 15.3 $E_R(n, j)/(m_e c^2)_{Codata} = \{f(n, j) - 1\}\,\{1 + \gamma\, m_e/M\}^{-1}\,\gamma$

All terms on the right-hand side of Relation 15.3 are dimensionless and calculable by number constants, if for α and m_e/M geometrized values and for γ the value of Definition 15.1 are used.

15.3 The 1S–2S transition in atomic hydrogen and the famous Lamb shift

Relation 15.3 allows to make a prediction for the most accurately measured frequency difference $h\Delta f(1S_{1/2}, 2S_{1/2}) = E_R(2,1/2) - E_R(1,1/2)$ of the hydrogenic atom. With $\alpha = \alpha_{geom}$, $m_e/M \approx 1/1{,}836.157672$ and γ from Definition 15.1, a value of **≈2466.0608 THz** for Δf is obtained. The deviation $\Delta f_{exp} - \Delta f_{theo}$ to the measured value of **≈2466.0614 THz** [CODATA-2014, Table X] is ≈0.6 GHz and is smaller than the hfs in the ground state of ≈1.42 GHz induced by the nuclear magnetic field of the proton. Has numerical coincidence helped achieve this outstanding result? With α_{Codata} and $\gamma = 1$, according to current doctrine, the deviation is about ten times larger, namely ≈−7.1 GHz. What is the cause of this rather massive discrepancy compared to the calculation with α_{geom} and γ?

The current doctrine explains the poor experimental agreement for α_{Codata} and $\gamma = 1$ by the postulate that for every state, in addition to the Dirac term $E_D(n, j)$ and the hyperfine energy $E_{hfs}(n, j, l, F)$, a so-called Lamb shift $L(n, l, j)$, first calculated by Hans Bethe using nonrelativistic QED, must be considered. Both the hyperfine energy and the Lamb shift depend on the orbital angular momentum l and allow to differentiate states with different orbital angular momenta. The Lamb shift, which has no quantum mechanical explanation, is primarily the result of small deviations of the Coulomb law at small distances, and the self-energy of the bound electron, that is, the virtual interaction with its own radiation field. The Lamb shift calculation is based entirely on singular mathematics and, in contrast to the hyperfine energy E_{hfs}, has no classical analogue.

The Lamb shift model was introduced in 1947 because, in contrast to the prediction of the Dirac approach, W. Lamb and R. Retherford measured a difference of $\approx$1 GHz between the energy levels $2S_{1/2}$ and $2P_{1/2}$. Since in the ground state no $1P_{1/2}$ state exists, in contrast to the $2S_{1/2}$ state, no $1S_{1/2}$ Lamb shift can be measured with radio frequency methods. The $1S_{1/2}$ Lamb shift must be determined by combining experimental data and QED calculations, which are basically undisputed today. As a result, "calculated" values cannot be interpreted as theoretical predictions because experimental data were already used to determine calculated values by means of least-square methods. Due to its high measurement accuracy, this is especially true for the frequency transition $\Delta f(1S_{1/2}, 2S_{1/2})$ of the hydrogenic atom. Does the simultaneous use of experiment and theory not contradict the basic idea that measurement and theory should be independent?

Since all excited states are unstable due to the emission of electromagnetic radiation, and consequently complicate both calculations and experiments, in what follows, only ionization energies shall be considered instead of 1S–2S transitions. Not only the stability of the states involved but also the fact that in the ground state short-range interactions are most pronounced justify an investigation of ionization energies. The interaction between the nucleus and the electron may also provide some insight into the fundamental problem why a bound, rotating charge, contrary to classical opinion, does not radiate in the ground state.

15.4 Ionization energies and the interaction of the electron with the nucleus

Unfortunately, there is no listing of experimental ionization energies or binding energies of nucleus-electron interactions available in the literature. However, a fairly large amount of precise measurement data of transition frequencies or combinations thereof [CODATA-2014, Table X] can be retrieved, which are used today to calculate ionization energies by means of a least-square calculation and the Lamb model. The following comparisons have therefore to be viewed on the premise that the iterative model used by NIST represents a suitable calculation rule with "correct" results, that is, it reliably reproduces the ionization energies of a system [15.1]. However, despite overall consistency the National Institute of Standards and Technology (NIST) model does not predict all experimentally measured transition frequencies with the same accuracy, that is, the formalism does not seem to be quite correct [15.3].

The ionization energy $I_{particle}$ or the maximum electron binding energy of a hydrogen-like system is theoretically determined by $E_R(1,1/2)$. For hydrogen, using α_{geom}, $m_e/M \approx 1/1{,}836.157672$ and γ given by Definition 15.1, $I_{theo} \equiv E_R(1,1/2)$ amounts to ≈13.598425 eV. The difference $I_{NIST} - I_{theo}$ to the tabulated NIST value of ≈13.598434005 eV is ≈ 2.115169 GHz, which is **1.9995** times the measured $2S_{1/2}$ Lamb shift $\Delta f(2P_{1/2}, 2S_{1/2})$ of ≈1.057845 GHz [CODATA-2014, Table X]. Why a deviation of this magnitude? Is the factor of nearly 2 just a numerical coincidence?

Is the Lamb shift of bound states truly the cause of vacuum fluctuations of free space, which modify the central field or the pure Coulomb potential at small distances, or could it be that the shift of energy levels is the result of a still unknown exchange process between electron and nucleus? In any case, bound states are physically completely different than unbound states, are even classically extremely complicated and can be understood only relativistically. Closely related to the foregoing is the question of whether bound states can be correctly described by QED or by quantum field theory, both of which cause mathematical singularities rooted in the concept of point particles and their interactions. In the following, an alternative is presented, which is applied for the ground state only. What the formalism for excited states might look like is unclear.

To interpret the difference between the measured value and what is predicted by Relation 15.3, the binding energy E_R shall be modified by a small dimensionless perturbation δE_R. More concretely,

Ansatz 15.1 $E/E_R = 1 + \delta E_R$ with $\delta E_R = (\mu/\mu_e)\,(q_e/q_m)^2$

introduces phenomenologically, in addition to the pure Coulomb field, a "magnetic binding mechanism" between the nucleus and the electron that scales with E_R.

Because of the definition $\mu_e/\mu_B \equiv g_e/2$ and the Dirac relation $q_e/q_m = 2\alpha$,

Relation 15.4 $E/E_R = 1 + \mu/\mu_B\,(2/g_e)\,(2\alpha)^2$

follows, where use has been made of Ansatz 15.1. The magnetic moment μ responsible for the binding (antibinding) contains the information about the interaction, that is, μ is a consequence of the interaction defined by $\mu/\mu_e\,(2\alpha)^2$ and does not represent the magnetic moment of the free nucleus μ_{nucl}.

If in the case of hydrogen μ/μ_B is equated to $2/3/\{\lambda_{e\_bar}/L_2/(L) + 5\}$, the hydrogenic ionization energy $I_H = E(1, 1/2)$ amounts to ≈13.598434027 eV, if α_{geom}, γ, $m_e/M \approx 1/1{,}836.157672$ and the g-factor of the free electron $g_e/2(\text{free}) = 1+a_e$ are also utilized. The deviation from the NIST value is ≈0.03 μeV, which is approximately 200 times smaller than the hfs of the ground state of ≈6 μeV. Surprisingly, the quantity $1/3/\{\lambda_{e\_bar}/L_2/(L) + 5\}$ or $\mu/2$, as explained in Section 14.6, roughly corresponds to the magnetic moment of the proton in units of μ_B.

Is it a coincidence that the electron binding energy of the hydrogenic ground state can be explained so simply with pure numbers and a small number of building blocks? Can ionization energies of other hydrogen-like systems be calculated by analogous correction terms? Can this potentially provide new insights into the interaction or the exchange process between the nucleus and the electron?

Table 15.1 lists calculated ionization energies of other hydrogen-like systems. As can be seen, simple terms for μ/μ_B allow to approximate phenomenologically ionization energies of deuterium (^2H), tritium (^3H) and other hydrogen-like systems similar to that of hydrogen (^1H). They reflect in an impressive way that energy levels somehow depend on nuclear structure and, without nuclear structure, basically no spectra can be precisely interpreted.

Table 15.1: Ionization energies of hydrogen-like systems.

	M/m_e	Z	μ/μ_B	I (cm^{-1})	I_{NIST} (cm^{-1})
^1H	1,836.152672	1	$+2/3/\{\lambda_{e\_bar}/L_2(L) + 5\}$	109,678.77196	109,678.77174
^2H	3,670.482966	1	$-1/3/\{\lambda_{e\_bar}/L_2(L) - 6\}$	109,708.61460	109,708.61455
^3H	5,496.921536	1	$-2/3/\{\lambda_{e\_bar}/L_2(L) - 2\}$	109,718.54382	109,718.54390
$^3He^+$	5,495.88528	2	$-8/5/\{\lambda_{e\_bar}/L_2(L) + 4\}$	438,889.190	438,889.194
$^4He^+$	7,294.29954	2	$-8/5/\{\lambda_{e\_bar}/L_2(L) - 2\}$	438,908.870	438,908.878
$^{12}C^{5+}$	21,868.6639	6	$-4/3/\{\lambda_{e\_bar}/L_2(L) + 4\}$	3,952,061.59	3,952,061.67
$^{16}O^{7+}$	29,148.9497	8	$-6/5/\{\lambda_{e\_bar}/L_2(L) + 8\}$	7,028,395.8	7,028,394.7

Note: Examples of ionization energies I = E(1,1/2) calculated with Relation 15.4. If the proton number Z is greater than 1, the fine-structure constant α must be replaced by (Zα) in the formula for both E and E_R. The ground state of tritium is unstable, in contrast to the ground state of all other particles, because this isotope is radioactive. The mass ratios of the nuclei $^1H^+$(proton) and $^2H^+$(deuteron) are geometrized values from Section 14.2. The mass ratios of the nuclei $^3H^+$(triton), $^3He^{2+}$(helion) and $^4He^{2+}$(alpha particle) are CODATA 2014 values. The mass ratios of the nuclei $^{12}C^{6+}$ and $^{16}O^{8+}$ were derived in analogy to Formula 15.6 using theoretically calculated ionization energies and atomic masses of NIST.

Also, the Lamb theory cannot sweep nuclear effects under the rug and must be extended by introducing a further free parameter. In addition to the hundreds of QED terms, the mean square nuclear charge radius, which requires corrections on the order of 1 MHz, is taken into account. But, as discussed extensively in the literature, the calculation of the charge radius from a comparison of theory and experiment of the Lamb shift leads to contradictions that have not yet been solved [15.2].

15.5 The hfs and the Fermi contact formula

For every complex problem there is an answer that is clear, simple, and wrong.
(*H. L. Mencken*)

The hfs of the ground state of hydrogen, its isotopes and $^3He^+$ is experimentally known with high accuracy. In all cases the hfs exists without an external magnetic field as well. The hydrogenic splitting is a few μeV, which leads to the emission or absorption of electromagnetic radiation at a wavelength of 21 cm. This subject was debated in detail in Section 6.2.

The dynamics of the interaction between the nucleus and the electron is very complicated, and the process is theoretically not understood in every detail because, strictly speaking, both the finite size, the internal structure and the polarizability of the nucleus must be taken into account. There is no doubt that the hfs represents the simplest and most fundamental magnetic interaction in atomic physics, and it is based on relativistic effects.

Especially for spherically symmetric states, the contact formula of E. Fermi, who derived it in 1930, plays a central role. It describes the interaction of the magnetic moment of the nucleus μ_{nucl} with the magnetic field generated by the electron inside the nucleus. Assuming that the angular momentum J of the electron and the angular momentum I of the nucleus are coupled by their magnetic moments and form a total angular momentum F = I + J, the hfs is given by

Formula 15.2 $$\Delta E_{hfs}\,(F-1, F)/(m_e c^2) = C\,\alpha^4\,(\mu_{nucl}/\mu_B)\,(g_e/2)\,(1 + m_e/M)^{-3}$$

where C represents a constant, dependent on the number of protons Z, the quantum numbers n, L, J, I and F, and is given by

Formula 15.3 $$C \equiv (Z/n)^3\,F\,\{\,J(J+1)\,(2L+1)\,I\,\}^{-1}$$

For the definitions of n, L, J, I, F and the derivation of the formulas, reference is made to the literature. A compact representation can be found, for example, in [15.3]. The factor $(1 + m_e/M)^{-3}$ takes into account the movement of the nucleus of mass M.

If the electron mass m_e is scaled by γ, that is, m_e is replaced by $\gamma\, m_{e\ Codata}$, then Formula 15.2 gives

Relation 15.5 $$\Delta E_{hfs}/(m_e c^2)_{Codata} = C\,\alpha^4\,(\mu_{nucl}/\mu_B)\,(g_e/2)\,(1+\gamma\, m_e/M)^{-3}\,\gamma^2$$

which depends on the six dimensionless parameters C, α, μ_{nucl}/μ_B, $g_e/2$, m_e/M and γ. In what follows various values calculated by Relation 15.5 are compared with experimental findings.

15.5.1 Hydrogen: the most accurate hfs measurement ever made

For atomic hydrogen in the ground state, that is, for $Z = 1$, $n = 1$, $L = 0$, $J = 1/2$, $I = 1/2$ and $F = 1$, Formula 15.3 gives $C_H = 8/3$. Since the electron has no orbital angular momentum L in this case, the total angular momentum $J = S + L$ is solely due to the spin S of the electron.

The empirical g-factor of the electron or the proton in the bound state, that is, when the environment of the particle is changed, is not identical to the g-factor of the free, isolated electron or proton. In other words, the magnetic moment of a free nucleus and the magnetic moment of a nucleus surrounded by electrons are not identical. Depending on the situation, a theoretical correction must be applied for each particle, which is difficult to determine. Theory and experiments of these problems are described in detail in [CODATA-1998, pp 378].

It has been found experimentally [15.4] that in the ground state of the hydrogen atom the bound g-factor of the electron is given by

Relation 15.6 $\quad g_e/2(\mathrm{H}, 1s) = g_e/2(\mathrm{free})\,(1–1.7709\cdot 10^{-5})$

because of its interaction with the proton. The binding correction $1.7709(13)\cdot 10^{-5}$ in Relation 15.6 corresponds, to a first approximation, to the so-called Breit term $(Z\alpha)^2/3$, which occurs because of the existing magnetic nuclear field [CODATA-1998, Formula 84]. This correction was derived in 1928 by Gregory Breit for a pure Coulomb potential and a point-like nucleus with proton number Z by solving the Dirac equation in the presence of an external magnetic field (Zeeman effect).

Surprisingly, nuclei such as the proton [CODATA-1998, Formula 87], the deuteron [CODATA-1998, Formula 90] and the positive muon (μ^+) in muonium [CODATA-1998, Formula 93] have a theoretical correction of the same order of magnitude. In [CODATA-2014, Table XIII] various theoretical ratios $g^{\mathrm{bound}}/g^{\mathrm{free}}$ are compiled.

Why does, besides the charge, neither the mass nor the spin of the nucleus significantly influence the bound g-factor? The current doctrine cannot reliably calculate nuclear magnetic moments because their origin is unknown. Despite this ignorance of physics at nuclear scales, binding corrections for the g-factors of the nuclei are determined very precisely. How is that possible?

Setting $\gamma = 1$, $\alpha = \alpha_{\mathrm{Codata}}$, $m_e/M \approx 1/1{,}836.15267$, $\mu_{\mathrm{prot}}(\mathrm{H})/\mu_B/10^{-3} \approx 1.5210322053$ [CODATA-2014], and using $g_e/2(\mathrm{H})$ from Relation 15.6, the hfs $(\Delta f)_{\mathrm{hfs}}(\mathrm{H})$ in the ground state amounts to ≈1.420460 GHz according to Relation 15.5. In the literature [15.5], a theoretical estimate of 1.4204031(8) GHz can be found, which involves all corrections (QED and proton structure). Despite all corrections, this estimate is too inaccurate for the CODATA group to play a role in the calculation of the natural constants. The pure QED estimate is ≈1.420452 GHz [15.6]. Due to "lack of knowledge" about the structure of the proton, the necessary nuclear correction terms cannot or only vaguely be calculated, and the high measurement accuracy of the hfs of atomic hydrogen is not used today for the determination of natural constants. One of the most accurate measurements remains untapped.

Using γ from Definition 15.1, α_{geom}, $\mu_{proton}(H)/\mu_B = 1/3/\{\lambda_{e\_bar}/L_2/(L) + 5\}$ and the same values for m_e/M and $g_e/2(H)$ as before, a value of ≈**1.4204052 GHz** is obtained, which only deviates ≈0.60 kHz from the measured value [15.6] of **1.420405751768(1) GHz.** The experimental value is based on hydrogen maser experiments, which have to be corrected due to different reasons, so that the frequency of the undisturbed transition $(\Delta f)_{hfs}(H, 1s)$ is obtained. If the electron g-factor $g_e/2(H)$ from Relation 15.5 is replaced by $g_e/2(H) = 1 + a_e(\text{free}) - 1/3\ \alpha_{geom}^2$, the deviation is reduced to ≈0.54 kHz (0.4 ppm). For a_e(free), the value from Ansatz 11.4 is to be used.

A simple model with few parameters gives an excellent agreement with the experiment. This is all the more remarkable as all the parameters used, except the additional electron binding correction $1/3\ \alpha_{geom}^2$, also occur in the calculation of the ionization energy of atomic hydrogen. Is it a coincidence that, even without a binding correction for the nuclear magnetic moment of the proton, such a good experimental agreement is achieved?

From $\mu_{proton}(H)/\mu_B = 1/3/\{\lambda_{e\_bar}/L_2/(L) + 5\}$ and $g_e/2(H) = 1 + a_e(\text{free}) - 1/3\ \alpha_{geom}^2$, the ratio $\mu_e(H)/\mu_p(H)$ can be calculated, if the identity $\mu_e \equiv g_e/2\ \mu_B$ is used. The numerical value of $\mu_e(H)/\mu_p(H)$ amounts to ≈658.2193, which can be compared with the literature value [CODATA-2014, Table XVIII, B30] of 658.210706(6), which was derived in 1972 from the spin-flip frequency of the proton and the electron of a hydrogen maser in a strong magnetic field [15.7]. The first measurement took place in 1966 and showed a value of 658.21049(20) [15.8]. Both experimental values are compatible with each other, but significantly deviate from the value calculated from $(\Delta f)_{hfs}(H)$ with γ and α_{geom}. Is the calculation of the ratio from $(\Delta f)_{hfs}(H)$ wrong, or could it be that the dynamics of the electron and the nucleus in the hydrogen maser – and thus the ratio $\mu_e(H)/\mu_p(H)$ – is slightly disturbed by the superimposed magnetic field?

Despite the very good agreement of $(\Delta f)_{hfs}(H, 1s)$ with the measured value, a discrepancy of 10 ppm to the literature value is obviously to be observed for $\mu_e(H)/\mu_p(H)$. This is puzzling because from $\mu_e(H)/\mu_p(H)$ by means of theoretical corrections [CODATA-2014, Table XXIV, B30] the ratio $\mu_e(\text{free})/\mu_p(\text{free})$ is calculated, which together with a_e serves for the indirect determination of $\mu_{proton}(\text{free})/\mu_B$, or together with m_μ/m_e for the determination of the anomaly a_μ according to Formula 15.4. In other words, the ratio $\mu_e(H)/\mu_p(H)$ combined with complex theoretical binding corrections also determines the anomaly a_μ. This context will be discussed again later in this chapter.[1]

1 For many years, the indirect method has provided the most accurate value for $\mu_{proton}(\text{free})/\mu_B$. Recent, direct investigations [15.9] of a single proton in a Penning trap yielded μ_p = 2.792847350(9) μ_N. This new value is consistent with the maser value, which is the tabulated CODATA 1998 value of

To introduce another binding correction $\delta g_e/2(H)$ for a better experimental agreement is certainly very speculative. Under the premise of simplicity, however, it may be possible to gain a better understanding. It is striking that with the simple assumption $\delta g_e/2(H) = \alpha_{geom}{}^3$ a value of **≈1.420405761 GHz** is obtained for $(\Delta f)_{hfs}(H, 1s)$. Thus, the relative deviation to the measured value only amounts to $\approx 7 \cdot 10^{-9}$.

The hfs of $2S_{1/2}$ has not been measured very often in recent years and is therefore experimentally not as well known as that of the ground state. Because of the $1/n^3$ dependence of the C-factor, it is about 1/8 times smaller and experimentally amounts to **0.177556860(16) GHz** [15.6].

Setting $C_H(2s) = 1/3$ and $g_e/2(H, 2s) = 1 + a_e(\text{free}) + 1/3\ \alpha_{geom}{}^2$, the theoretical value of **≈0.177556925 GHz** is obtained with a deviation of ≈65 Hz (0.4 ppm) to the measured value. All other parameters are the same as in the calculation of the hfs of the ground state. It is noteworthy that only the inverse of the sign in the binding correction of the electron g-factor leads to an excellent agreement of 0.4 ppm with the measured value, and the correctness of the theoretical assumption $\mu_p(H)/\mu_B = 1/3/\{\lambda_{e\_bar}/L_2/(L) + 5\}$ seems to be confirmed, if no binding correction is additionally used for the nuclear magnetic moment of the proton.

The hfs of $2S_{1/2}$ was measured exclusively by optical methods and without a disturbing, external magnetic field. If, in analogy to $\delta g_e/2(H, 1s)$, an additional correction $\delta g_e/2(H, 2s) = -\alpha_{geom}{}^3$ is introduced, a value of **≈0.177556857 GHz** results for the hfs of $2S_{1/2}$. Is this good agreement with the experiment underpinning the correctness of the speculative assumption of $\delta g_e/2(H, 1s)$?

15.5.2 Muonic hydrogen (μp)

The validity of the scaling of Relation 15.5 by γ can be "checked" with the aid of muonic hydrogen. In this case, the factor γ is ≈206.768 due to the larger mass of the muon compared to the mass of the electron. Because the proton is still the nucleus, the ratio m_e/M does not change and amounts to ≈1/1,836.152672 as for atomic hydrogen. Because the electron–muon universality does not hold, the electron g-factor of atomic hydrogen $g_e/2(H)$ must be replaced by the muon g-factor of muonic hydrogen $g_\mu/2(\mu p)$.

The splitting of the $2S_{1/2}$ state in muonic hydrogen is experimentally known as a differential measurement of two Lamb transitions. With the help of calculated values of the 2P fine structure and the $2P_{3/2}$ hyperfine structure, A. Antognini and

---

$\mu_p(\text{free}) = 2.792847337(29)\mu_N$ or $\mu_p(\text{free})/\mu_B/10^{-3} = 1.521032203(15)$, respectively. In the year 1969, the CODATA group compiled the value 1.52103178(23) for $\mu_p(\text{free})/\mu_B/10^{-3}$. All theoretical assumptions and complicated corrections applied for the indirect method must therefore be correct. Everything is straight. But, it could also be that the required systematic corrections for the direct method fit so well that both methods agree without contradiction. What if they would contradict each other?

coworkers experimentally obtained **22.8089(51) meV** [15.10]. This value is not completely free from theoretical assumptions. Since all parameters of the muonic C-factor are the same as for hydrogen, the $C_{\mu p}(2s)$ factor of the $2S_{1/2}$ state of muonic hydrogen is 1/3.

Using $\gamma = (m_\mu/m_e)_{geom}$, α_{geom}, $\mu_{proton}(\mu p)/\mu_B = 1/3/\{\lambda_{e\_bar}/L_2/(L) + 5\}$ and the muon g-factor $g_\mu/2(\mu p, 2s) = g_\mu/2(\text{free}) = 1 + a_\mu$, the result for $(\Delta f)_{hfs}(\mu p)$ in the $2S_{1/2}$ state is ≈22.6681 meV, which does not quite correspond to the literature value [15.10]. Surprisingly, however, the assumption $g_\mu/2(\mu p, 2s) = 1 + \alpha_{geom}$ excellently matches the literature value within its error limits, namely, the calculation yields **≈22.8067 meV**.

The question is naturally raised, why such a good agreement arises when the Schwinger correction $\alpha_{geom}/(2\pi)$, which occurs in the anomaly a_μ as the largest term and stands for the interaction with the virtual radiation field, is simply replaced by α_{geom}. This means, in other words, that $g_\mu/2(\text{free})$ cannot be obtained from the bound muon g-factor $g_\mu/2(\mu p, 2s)$ as a limiting process of vanishing nuclear charge.

Is it all a coincidence, or is it attributable to the fact, that the energy scale of the muon is 206 times the energy scale of the electron, and therefore a muonic binding interaction arises, which is proportional to the fine-structure constant? Perhaps, a muon near the proton behaves differently than an electron, and the important lepton universality of the Standard Model needs to be reconsidered. Because of its mass, the muon spends more time near the proton, and it may be that the muon interacts with localized constituents of the nucleus differently than an electron further away from the nucleus. This is also linked to the famous, until now unexplained question of I. I. Rabi, why the muon as a particle only exists for a short time, even though it has similar properties as the electron. Besides cataloguing the muon as a lepton, the current doctrine is silent on what else it might be.

15.5.3 Deuterium

The atomic deuterium has a weakly bound, easily polarizable nucleus with I = 1. The remaining quantum numbers are the same as those of atomic hydrogen. The nucleus (deuteron) has a quadrupole moment, which in addition to the Fermi-contact interaction leads to another, but very weak interaction. In the following, this interaction will be neglected and only the simple Fermi contact interaction will be considered.

Using $C_D(1s) = 2$, γ, α_{geom}, $\mu_{deuteron}(D)/\mu_B = 1/10/\{\lambda_{e\_bar}/L_2/(L)\}$ and the bound electron g-factor $g_e/2(D) = 1 + a_e + 10/3\ \alpha_{geom}^2$ the result for the hfs $(\Delta f)_{hfs}(D)$ in the ground state is **≈0.327384425 GHz**, which deviates about 73 Hz from the measured value of **0.327384352522(2) GHz** [15.6]. For m_e/M the reciprocal mass of the deuteron from Table 15.1 should be used. The nuclear magnetic moment of the deuteron $\mu_{deuteron}(D)$

in units of μ_B corresponds to the nuclear magnetic moment listed in Table 14.2 without additional binding corrections.

The pure QED estimate is ≈0.327339 GHz [15.6]. The large difference of about 45 kHz to the measured value is solely attributed to nuclear effects, whose calculations are still very controversial today because of lack of knowledge. Since the theoretical uncertainty of today's doctrine for deuterium is even greater than that for hydrogen, the hfs $(\Delta f)_{hfs}(D, 1s)$ plays no role whatsoever in determining CODATA values. Another precise measurement is left unused for many years.

Why does the parameterization with simple natural numbers achieve a relative deviation of $\approx 2{\cdot}10^{-7}$, and why does the same factor 10 occur in the electron binding correction $10/3\ \alpha_{geom}^2$ as in the nuclear magnetic moment $\mu_{deuteron}(D)$? Is this an indication that the electron senses the properties of the deuteron structure? The factor $1/3\ \alpha_{geom}^2$ is present in the electron binding correction of $(\Delta f)_{hfs}(H, 1s)$, $(\Delta f)_{hfs}(H, 2s)$ and $(\Delta f)_{hfs}(D, 1s)$. Is everything just coincidence or can valuable information for a different understanding be gained from this fact?

The good relative agreement of ≈0.2 ppm for $(\Delta f)_{hfs}(D, 1s)$ is being dampened by the fact that the ratio $\mu_d(D)/\mu_e(D)$ of $\approx 4.6633{\cdot}10^{-4}$ deviates about 250 ppm from the literature value of $4.664345392(50){\cdot}10^{-4}$ [CODATA-1998, Formula 100]. The literature value was determined in 1984 by MIT staff members using the same method as for $\mu_e(H)/\mu_p(H)$ [15.7], but was never published. There are no competitive values and the measured ratio is therefore still used by the CODATA group as an input parameter [CODATA-2014, Table XVIII, B31] for the determination of $\mu_{deuteron}(\text{free})/\mu_B$. Why is there an experimental deviation for $\mu_d(D)/\mu_e(D)$, which is more than a factor of ten larger than that observed for $\mu_e(H)/\mu_p(H)$? Is the maser method, due to the existing strong magnetic field, for the determination of $\mu_d(D)/\mu_e(D)$ more likely to be affected by systematic effects than in the case of hydrogen?

The additional correction $\delta g_e/2(D, 1s) = -4/7\ \alpha_{geom}^3$ implies a theoretical value of **≈0.327384352884 GHz** for the hfs $(\Delta f)_{hfs}(D)$ in the ground state. The relative deviation to the measured value, which is based on deuterium maser experiments analogously to hydrogen, is reduced with this correction to $\approx 10^{-9}$. It is remarkably small, even seven times smaller than for hydrogen. The ratio 4/7 or its reciprocal is not unknown in this book. It crops up, for example, in the parameterization of the mean nuclear binding energy (Relation 8.6) and in the description of the neutron decay (Section 14.4). In both cases, the ratio has a relation to the neutron. Since the fraction 4/7 is already important in other contexts, no new free parameter is introduced by the correction $\delta g_e/2(D, 1s)$. In spite of the excellent agreement with experiment, the connection to the neutron, the simplicity and the use of simple natural numbers, there will probably be critics who see the small deviation as pure coincidence.

Similar to hydrogen, there are also measurements of the hfs of the deuterium $2S_{1/2}$ state. In this case, the C-factor $C_D(2s)$ is 1/4. The value of $(\Delta f)_{hfs}(D, 2s)$ measured with classical radio frequency methods amounts to 40,924.439(20) kHz [15.11]. Recently,

corrections, is estimated by the CODATA group to be 4.463302891(272) GHz [CODATA-2010, Formula 221], comprising the value calculated by Relation 15.5 with γ and α_{geom}.

Since the virtual contribution $4/3\ \alpha_F/\alpha_{geom}$, introduced in Section 11.4 for provisional agreement with the literature value of the anomaly a_μ, increases the deviation from the measurement for $(\Delta f)_{hfs}(Mu)$, this term was omitted in previous calculations. Therefore, the value used for the anomaly a_μ, unlike a_e, imprecisely reflects the tabulated measurement of the free magnetic moment of the muon. Why is there this asymmetry for the muon in contrast to the electron?

Contrary to the electron magnetic moment anomaly a_e, the anomaly a_μ is not determined directly, but deduced from the experimental quantity R using

Formula 15.4 $\quad R = a_\mu\ (m_\mu/m_e)^{-1}\ \{\mu_{proton}(free)/\mu_B\}^{-1} = 0.0037072063(20)$

[CODATA-2014, Formulas 135 and 137] because the magnetic field is determined from nuclear magnetic resonance measurements. The value R represents a ratio of measured frequencies, is dimensionless and independent of models. Using CODATA values for m_μ/m_e and $\mu_{proton}(free)/\mu_B$, it can easily be verified that, according to Formula 15.4, the anomaly a_μ amounts to ≈0.001165921, which is recorded in the literature today.

Without doubt, Formula 15.4 can also be used to deduce the magnetic moment of the free proton in units of μ_B when use is made of theoretical approaches for the anomaly a_μ and the mass ratio m_μ/m_e. With $(m_\mu/m_e)_{geom}$, according to Ansatz 14.2, and the anomaly $a_\mu = \alpha_{geom}/(2\pi) + \{\alpha_F/(2\pi)\}^{1/2}$, which was also included in the calculation of $(\Delta f)_{hfs}(Mu)$, the magnetic moment of the free proton $\mu_p(free)/\mu_B/10^{-3}$ amounts to ≈1.520968. Interestingly, this value deviates only about 12 ppm from $1/3/\{\lambda_{e\_bar}/L_2/(L) + 5\}/10^{-3}$ or ≈1.520985, respectively. Why does such a self-consistent picture emerge from a small set of parameters, which are all based on mathematical constants, and why can the dimensionless quantity $1/3/\{\lambda_{e\_bar}/L_2/(L) + 5\}$, without complicated corrections, excellently describe many experimental phenomena involving a single proton?

Assuming that the magnetic moment of the free proton in units of μ_B can be equated to $1/3/\{\lambda_{e\_bar}/L_2/(L) + 5\}$, Formula 15.4 yields for $a_\mu = \alpha_{geom}/(2\pi) + \{\alpha_F/(2\pi)\}^{1/2}$ a correction δa_μ of $\approx\alpha_F/\alpha_{geom}/3$. The anomaly $a_\mu = \alpha_{geom}/(2\pi) + \{\alpha_F/(2\pi)\}^{1/2} + \alpha_F/\alpha_{geom}/3$ has a value of ≈0.001165883 and corresponds quite well to the value of ≈0.001165884 calculated from Formula 15.4. With this correction, however, for $(\Delta f)_{hfs}(Mu)$ the deviation from the measured value [15.14] increases from 0.08 to 0.1 ppm, but which is still within the error band of the measured value determined with a weak magnetic field [15.16]. Both 1/3 and α_F/α_{geom} occur in Ansatz 14.2 for m_μ/m_e, and thus support the introduction of a correction term as described previously.

Of course, it has to be clarified for the muonium which physical meaning the symmetric binding correction has or which relativistic interaction between the electron and the muon causes a magnetic correction of the type $7/5\ \alpha_{geom}^2$ for both the electron and the muon. Why is the electron correction in hydrogen $1/3\ \alpha_{geom}{}^2$? It is interesting that the two primes 5 and 7 also arise in Ansatz 14.2 for the muon mass in units of m_e and in Definition 15.1 of the Lorentz-invariant scaling factor $\gamma = \{1 - (4{\cdot}5{\cdot}7)^{-1}\}^{-1/2}$.

Is all this a coincidence and just happened by playing with mathematics, or could it be that, as already suggested in Section 3.2, the nine-dimensional hypersphere is the key to a unified understanding and around which nature is elegantly organized? In all the experimental comparisons in this book, only the numbers 2, 3, $4(=2^2)$, 5, $6(=2{\cdot}3)$, 7, $8(=2^3)$, $9(=3^3)$ or $10(=2{\cdot}5)$ were used in the mathematical expressions in addition to the number π. Only in a few cases, there is a reference to the abstract concept of the real number.

In order to confirm this fact, all the terms used in Relation 15.5 for g/2, and μ_{nucl} in units of the Bohr magneton μ_B, are briefly summarized below:

$g_e/2(\text{H, 1s}) = 1 + a_e - 1/3\ \alpha_{geom}{}^2 + \alpha_{geom}{}^3$
$g_e/2(\text{H, 2s}) = 1 + a_e + 1/3\ \alpha_{geom}{}^2 - \alpha_{geom}{}^3$
$g_\mu/2(\mu\text{p, 2s}) = 1 + \alpha_{geom}$
$g_e/2(\text{D, 1s}) = 1 + a_e + 10/3\ \alpha_{geom}{}^2 - 4/7\ \alpha_{geom}{}^3$
$g_e/2(\text{D, 2s}) = 1 + a_e + 4\ \alpha_{geom}{}^2$
$g_\mu/2(\mu\text{d, 2s}) = 1 + 4\ \alpha_{geom}$
$g_e/2(\text{Mu}) = 1 + a_e - 7/5\ \alpha_{geom}{}^2$

$\mu_{proton}(\text{H}) = \mu_{proton}(\mu\text{p}) = 1/3\ /\ \{\lambda_{e\_bar}/L_2/(L) + 5\}$
$\mu_{deuteron}(\text{D}) = \mu_{deuteron}(\mu\text{d}) = 1/10\ /\ \{\lambda_{e\_bar}/L_2/(L)\}$
$\mu_\mu(\text{Mu}) = (m_\mu/m_e)^{-1}\ \{1 + a_\mu - 7/5\ \alpha_{geom}{}^2\}$
with $a_\mu = \alpha_{geom}/(2\pi) + \{\alpha_F/(2\pi)\}^{1/2} + 1/3\ \alpha_F/\alpha_{geom}$

15.6 The spin g-factor of the electron in $^{12}C^{5+}$, $^{16}O^{7+}$ and $^{28}Si^{13+}$

The particles $^{12}C^{5+}$, $^{16}O^{7+}$ and $^{28}Si^{13+}$, whose nuclei have no spin, are hydrogen-like ions with the proton number Z = 6, Z = 8 or Z = 14, respectively. In these particles, a single electron is bound to a spinless nucleus by a strong Coulomb field. Such ions can be confined or localized in a small spatial area in a cooled Penning trap by means of a strong, constant magnetic field and a weak electrostatic quadrupolar field. If the confined ion is additionally irradiated with microwaves, spin flips are induced, from which at the maximum spin flip rate the spin precession frequency f_s can be inferred. Together with the cyclotron frequency f_c of the ion, the spin g-factor g_e of the bound electron can then be determined. From the ratio $\Gamma \equiv f_s/f_c$ and with the help of

Formula 15.5 $g_e/2 = \Gamma\ (m_{ion}/m_e)^{-1}\ (Q_{ion}/q_e)$

the spin g-factor can be calculated, if the mass ratio m_{ion}/m_e is known. The charge ratio Q_{ion}/q_e is Z–1 and thus always whole-numbered. The cyclotron frequency f_c serves to calibrate the magnetic field acting on the electron.

The ionic masses of $^{12}C^{5+}$, $^{16}O^{7+}$ and $^{28}Si^{13+}$ can be calculated via the sum of their ionization energies. For example, the mass of $^{16}O^{7+}$ expressed in atomic mass units can be obtained from

Formula 15.6 $$m(^{16}O^{7+})/u \equiv A_r(^{16}O^{7+}) = A_r(^{16}O) - 7A_r(e) + E_{bind}/(uc^2)$$

where $A_r(^{16}O)$ is the mass of the neutral isotope ^{16}O, $A_r(e)$ is the mass of the electron in units of u and E_{bind} is the sum of the ionization energies of the first seven electrons. By definition, the mass ratio $m(^{16}O^{7+})/m_e$ agrees with $A_r(^{16}O^{7+})/A_r(e)$. Analogous formulas apply to $^{12}C^{5+}$ and $^{28}Si^{13+}$.

With the experimental data

$\Gamma(^{12}C^{5+})$	= 4,376.21050087(12)	[CODATA-2014, table XVIII, B15]
$\Gamma(^{16}O^{7+})$	= 4,164.3761837(32)	[CODATA-2010, table 20, B18]
$\Gamma(^{28}Si^{13+})$	= 3,912.86606484(19)	[CODATA-2014, table XVIII, B18]

the compiled NIST values [queried on September 30, 2017] of the atomic masses, the ionization energies, Ar (e) and u, we obtain from Formulas 15.5 and 15.6

$g_e(^{12}C^{5+})$	**≈2.001041590**	[≈2.001041590, CODATA-2006, table XV]
$g_e(^{16}O^{7+})$	**≈2.000047019**	[≈2.000047020, CODATA-2006, table XVI]
$g_e(^{28}Si^{13+})$	**≈1.995348958**	[≈1.995348958, CODATA-2014, table XII]

In the square bracket the QED value with the corresponding reference is cited. Despite theoretical ignorance of the nuclear structures, the experimental and theoretical figures are in all cases in excellent agreement with one another. It even seems to prevail almost complete agreement. Quite amazing. Since all QED calculations were known before the experiments, given such an astonishing conformity, the question arises to what extent there is a mutual dependence between theory and experiment.

Since ion masses can be measured more accurately than the relatively tiny electron mass, m_e is the parameter that limits the accuracy in Formula 15.5. As a result, nowadays Formula 15.5 is not used to deduce the factor g_e, but to indirectly optimize the mass m_e by calculating the electron g-factor with QED. The scientific benefits are additional decimal places of m_e, but with the mostly unmentioned fact that the mass m_e becomes even more model-dependent and can only be understood in association with QED.[3] Independence

3 As a first approximation, the mass $m_{e,\,Codata}$ historically depends on the Rydberg constant R_∞ of the imperfect Bohr model describing reality incompletely. QED calculations play a role at higher accuracy only.

from theory and measurement is different. Today, comparisons between theory and experiment are usually not used to validate bound state QED, but rather serve to more accurately determine fundamental constants or to study nuclear effects. QED with the problem of infinite mass and charge, that is, with its mathematically undefined subtractions remains sacrosanct for lack of alternatives.

Surprisingly, the experimental data for $^{12}C^{5+}$ and $^{28}Si^{13+}$ can also be parameterized with α_{geom} and Hall fractions. Numerical testing yields

$$g_e/2(^{12}C^{5+}) \quad = 1 + a_e - (1/3)(6\alpha_{geom})^2 - (7/4)(1/3)(6\alpha_{geom})^4 \quad \approx \mathbf{2.001041589}/2$$

$$g_e/2(^{28}Si^{13+}) \quad = 1 + a_e - (1/3)(14\alpha_{geom})^2 - (2/5)(3/7)(14\alpha_{geom})^4 \quad \approx \mathbf{1.995348948}/2$$

For a_e, the value of Ansatz 11.4 was used ensuring that only number constants are employed in the parameterization. The relative deviation $\delta g/g$ to the value calculated with Formula 15.5 is $6{\cdot}10^{-10}$ for $^{12}C^{5+}$, or $5{\cdot}10^{-9}$ for $^{28}Si^{13+}$, respectively, indicating that the point of view contains a great deal of truth. It is unclear why the $^{16}O^{7+}$ parameterization does not work so well. The best match results with

$$g_e/2(^{16}O^{7+}) \quad = 1 + a_e - (1/3)(8\alpha_{geom})^2 - (3/5)(3/5)(8\alpha_{geom})^4 \quad \approx \mathbf{2.000047075}/2$$

In the latter case the relative deviation amounts to $3{\cdot}10^{-8}$. Have $^{12}C^{5+}$ or $^{28}Si^{13+}$received some help by serendipity? Why is the deviation the smallest for $^{12}C^{5+}$, where the ion mass is most accurately known, since $A_r(^{12}C)$ is exactly 12 by definition?

References

[15.1] National Institute of Standards and Technology (NIST). Atomic Spectra Database Ionization Energies Form. http://physics.nist.gov/PhysRefData/ASD/ionEnergy.html

[15.2] Richard J. Hill. Review of experimental and theoretical status of the proton radius puzzle. EPJ Web of Conferences 137(2017), 01023. https://arxiv.org/abs/1702.01189

[15.3] Alexander E. Kramida. A critical compilation of experimental data on spectral lines and energy levels of hydrogen, deuterium, and tritium. Atomic Data and Nuclear Data Tables, 96(2010), 586–644

[15.4] J. S. Tiedeman and H. G. Robinson. Determination of $g_J(^1H, 1^2S_{1/2})/g_s(e)$: test of mass-independent corrections. Physical Review Letters 39(1977), 602–604

[15.5] Michael I. Eides, Howard Grotch and Valery A. Shelyuto. Theory of light hydrogenic bound states. Springer-Verlag, Berlin, 2007, table 11.1, p. 231

[15.6] Savely G. Karshenboim and Vladimir G. Ivanov. Hyperfine structure in hydrogen and helium ion. Physics Letters B 524(2002), 259–264. https://arxiv.org/abs/hep-ph/0109128

[15.7] P. F. Winkler, D. Kleppner, T. Myint and F. G. Walther. Magnetic moment of the proton in Bohr magnetons. Physical Review A 5(1972), 83–114

[15.8] Than Myint, Daniel Kleppner, Norman F. Ramsey and Hugh G. Robinson. Absolute value of the proton g-factor. Physical Review Letters 17(1966), 405–407

[15.9] A. Mooser, S. Ulmer, K. Blaum et al. Direct high-precision measurement of the magnetic moment of the proton. Nature 509(2014), 596–599

[15.10] A. Antognini, F. Nez, K. Schuhmann et al. Proton structure from the measurement of 2S-2P transition frequencies of muonic hydrogen. Science 339(2013), 417–420. https://doc.rero.ch/record/32524/files/kno_lsm.pdf

[15.11] H.A. Reich, J.W. Heberle and P. Kusch. Hyperfine structure of the metastable deuterium atom. Physical Review 104(1956), 1585

[15.12] N. Kolachevsky, P. Fendel, S. G. Karshenboim and T. W. Hänsch. 2S hyperfine structure of atomic deuterium. Physical Review A 70(2004), 062503. https://arxiv.org/pdf/physics/0405058.pdf

[15.13] R. Pohl, F. Nez, L. M. Fernandes et al. Laser spectroscopy of muonic deuterium. Science 353(2016), 669–673

[15.14] W. Liu, M. G. Boshier, S. Dhawan et al. High precision measurements of the ground state hyperfine structure interval of muonium and of the muon magnetic moment. Physical Review Letters 82(1999), 711

[15.15] G. Breit and I. I. Rabi. Measurement of nuclear spin. Physical Review 38(1931), 2082

[15.16] D. E. Casperson, T. W. Crane, V. W. Hughes et al. A new high precision measurement of the muonium hyperfine structure interval. Physical Letters B 59(1975), 397–400

Part II

16 The boundary between semimetal and insulator

Education does not come from reading but from thinking about what has been read.
(Carl Hilty)

16.1 Electrical transport measurements of thin film resistive layers with mesoscopic geometries

Materials that are – in comparison to the lateral extension – thinly condensed on a carrier substrate, such as glass or silicon, are called thin films. The resistive materials presented in this chapter belong to this category because they are only a few tens of nanometers thin. To electrically characterize these resistive materials, they must additionally be provided with a conductive layer for contacting. If both layers are independent of one another laterally tailored, so-called thin-film resistors are produced. By varying the material composition (sheet resistance) and by lateral, microscale structuring, with such layer stacks it is possible to build a variety of systems that have a spatial extension on the order of magnitude of the volume cell $\Lambda\, L\, \lambda_{Lo}$ with $\Lambda \approx 4.3\ \mu m$, $L \approx 12.7\ \mu m$ and $\lambda_{Lo} \approx 21$ nm, discussed in Section 5.2. The sheet resistance was adjusted by variable doping while maintaining the nanoscale layer thickness (20–50 nm) and the amorphousness, which was diffractometrically examined in certain systems.

It is astonishing that in the case of high-resistive[1] layers, that is to say in the case of material compositions with specific resistances that are typical for semimetals, both the sheet resistance and the temperature coefficient can depend on the geometry of the resistors. As a result, for small geometries, it does matter which area is used to calculate the sheet resistance. The fact that the geometry dependence must be more pronounced, the narrower or thinner the resistors are, is actually understandable. For infinitely narrow or thin geometries, the sheet resistance is no longer defined, since in the limiting case the cross section of the resistor becomes zero. It is worth noting, however, that even in lateral structures of the micrometer scale on the order of $\Lambda(\approx 4.3\ \mu m)$ and $L(\approx 12.7\ \mu m)$ unexpectedly sudden changes in resistance occur and Ohm's law collapses.

1 Layers with sheet resistances greater than 300 Ω are referred to as high ohmic. Their electrical properties are very similar to those of semimetals.

https://doi.org/10.1515/9783110612387-016

In the literature, it is little known about geometry dependencies (size effects) of thin film resistors. On the basis of experimental findings, a complexity is seemingly coming into play with high-ohmic, mesoscopic layers, which is unknown for solid materials and only noticeable in much smaller geometries of low-ohmic (metallic) systems. High-ohmic systems convey the impression that geometrical properties must already be important on the micrometer scale, since spatial influences are so clearly observable in the behavior of resistors. This chapter attempts, as an alternative to current doctrine, to shed some light on such effects using the scales τ, Λ, L and λ_{Lo} of previous chapters.

The similarity in temperature curves between –70 °C and +160 °C from many resistance measurements reflects the fact that within this temperature range similarity laws must be effective regardless of the material composition and the geometry. In this chapter it is shown that, despite the enormous diversity, a common behavioral basis can be found if the measurement data are normalized by the reference temperature τ of Section 5.2. This calibration reveals that indeed a heat bath of temperature τ on the order of ≈1,134 K could play a fundamental role.

At the end of this chapter, several observations are listed, which at first sight might be considered as "measurement anomalies". However, it is also possible that charge carriers in the material exhibited a collective behavior at the macrolevel that led to these anomalies in the measurement, or that the electrons were somehow coupled to "background fields". In any case, these measurements raise many questions that are not clear. They hold great secrets, as they cannot at all be explained with the models available today. What is the cause of why systems unexpectedly undergo non-reproducible phase changes with a small change in temperature? Unfortunately, this question is not completely established even if we take into account considerations of previous chapters.

It may seem daring to assume a connection between electrical transport behavior and fundamental scales of previous chapters. However, many observations provide enough evidence to confirm that fundamental principles of nature may possibly be tested by means of electrical transport measurements of charge carriers in an amorphous matrix. The extrapolation of the observations made with resistive layers of mesoscopic dimensions to fundamental questions of physics is risky. It is up to the reader to decide to what extent resistance measurements are suitable experiments to elucidate certain fundamental questions of physics, or whether they might give us more insight on the understanding of elementary quantum mechanical effects. Why should such measurements not be a "laboratory" that was hitherto reserved for high-energy physics?

At least among theoreticians, particle and solid-state physics are of a similar nature in their principles because the ion-lattice forming the solid can be regarded as an analog to the vacuum medium of a field theory. Probably the most famous concept, based on ideas of solid-state physics, is the Anderson–Higgs mechanism for the creation of particle masses. Now, this mechanism is a big topic of theoretical physics (symmetry breaking) and has become indispensable in today's formulation of modern interaction theories between particles.

All measurements were carried out before the elaboration of the ideas presented in the first part of the book in a different context in the company Cicor Micro Electronics Reinhardt Microtech AG. The choice of geometries and sheet resistances is not based on theory, but reflects other requirements. For reasons of secrecy, the characterization of the layers is deficient, that is, the physical properties are only explained so far that references to the ideas of the first part of the book can plausibly be drawn.

16.1.1 Short description of the samples used and the measuring principle applied

Electrical transport measurements provide important information about the behavior of charge carriers in materials. Experimentally, the electrical resistance value ($R_{meas}= U/I$) is measured as the ratio of the resulting (measured) voltage (U) to the impressed current (I). The inverse of R_{meas} is also called electrical conductance. The resistance value or the conductance should not be confused with the intensive quantities specific resistance (volume resistance or resistivity) and specific conductance, which for macroscopic samples are almost independent of the system size. These macroscopic characteristics are subject to units, are material dependent and do not constitute dimensionless constants.

For two-dimensional resistance systems, it is a common practice to consider the film thickness as constant and to specify the surface or sheet resistance of the film in units of ohm per unit area. The resistance is then determined as the number of unit squares in series. High-resistance values are achieved either by a material with a high sheet resistance and/or by a large number of squares in series. In the production of two-dimensional resistance systems, the last method is usually applied and the geometrical dimensions are reduced to the limit of the lithographic process to achieve a large number of squares. The scaling of the sheet resistance by changing the film thickness or the material composition by doping has similar practical limitations as the lithography. In any case, the interaction between film thickness and material composition (doping) is most interesting for materials science.

All measurements were made on samples that consisted of an insulating carrier substrate of dimension 50 mm × 50 mm, on which eight resistors were patterned into an area smaller than 4 mm × 1 mm. The size of the magnetron sputtering plant ensured that within this small area, the chemical composition and the interior structure of the amorphous layers can be considered as very homogeneous.

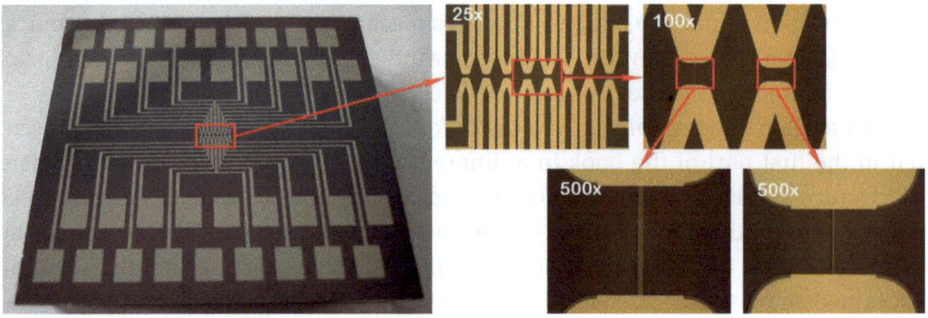

Figure 16.1: Sample geometry.
Note: A typical 50 mm × 50 mm test sample with eight patterned resistors of amorphous layers with various magnifications of some details. On the left side, the entire substrate with all macrocontacts (charge reservoirs) for contacting with spring contact probes is shown. The first section illustrates all eight resistors within the range of 4 mm × 0.2 mm used as the measuring area. The remaining magnifications are exemplary representations of two resistors down to the last detail. The cathode sputtering of the resistance material was carried out at room temperature. During the sputtering process, all substrates were rotated transversally to the contacts, that is, along the longer side of the resistor area, to improve homogeneity. Data provided with kind permission of Cicor Micro Electronics Reinhardt Microtech AG.

Each resistor was connected to two supply lines and could thus be measured in a four-contact arrangement. In a four-contact arrangement, two contacts serve as current contacts and the remaining two contacts enable an accurate measurement of the voltage drop over the resistor by means of a high-input impedance voltmeter. By applying a constant current in longitudinal direction and measuring the voltage drop caused thereby allow to calculate the resistance value as the ratio of the measured voltage to the impressed current. The extension of the resistor in the transverse direction of the impressed current is referred to as the resistor width, and the extension of the resistor in the direction of the impressed current is named the resistor length. The dimension perpendicular to the resistive surface is characterized by the film thickness.

16.2 The temperature dependence of the electrical resistance

Numerous experiments have been carried out to understand the conduction mechanism of amorphous materials as a function of temperature, and many empirical models for the temperature profile have been evaluated to fit the experiment. It is noteworthy that not only monotonically decreasing or increasing temperature profiles, but also curves with a minimum [16.1–16.4] and/or a maximum [16.5], [16.6] are known. These extremes are found within wide temperature ranges and can often be shifted in position by heat treatment.

Today, at low temperatures, resistance minima are theoretically explained by the Kondo effect, which attributes the anomalous behavior of metals to the scattering of conduction electrons at magnetic impurities. For resistance extremes at room temperature, however, there are no analogous theories available. These anomalies are in stark contradiction to the current theory of the electrical conductivity of metals. However, extremes in resistance curves in the range of room temperature are an integral part of high-ohmic materials, that is, in the boundary region between semimetal and insulator. Something fundamental seems to be missing in existing models of the electrical conductivity in the technically important temperature range. Birjega and Rau [16.7] even observed an oscillating behavior between 25 °C and 85 °C, which depended on the thermal history of the sample.

It is experimentally observed that the shape of the curve of the temperature dependence, regardless of the chemical composition or geometry of the resistor, behaves according to a certain pattern. Searching for a simple, universal behavior is therefore obvious and very valuable, so that a deeper understanding of the behavior of such systems can be gained. Despite the enormous complexity of matter in the microcosm, many experimental variables follow universal laws with few macroscopic parameters, if "correct" scaling is employed.

According to the phenomenological Matthiessen rule (scattering hypothesis), the specific resistance is composed of a temperature-dependent and a temperature-independent term. The temperature-independent term is referred to as specific residual resistance. In the Drude theory, all microscopic processes are integrated in a mean free path (or scattering time) and the different, independent scattering mechanisms are distinguished by different temperature dependencies. The inverse, mean free path lengths are each proportional to resistance, and the total resistance thus follows from the sum of the contributions of the independent scattering mechanisms.

To describe the residual resistance and the different temperature dependences of the independent scattering mechanisms in the sense of the previously discussed rules, the universal scaling law according to

Ansatz 16.1 $$R_{meas}(T) = a_2\,T^2 + a_1\,T + a_0 + \text{remainder}(T)$$

was chosen for fitting measured resistance curves as a function of temperature. The remainder shall take into account the deviation from a purely quadratic fit.

The assumption that the observable $R_{meas}(T)$ can be roughly described within a limiting temperature range by three power functions (T^0, T^1, T^2) and a small remainder is confirmed by many evaluations of experimental data of amorphous, high-ohmic materials. For the parameters a_1 and a_2, however, both positive and negative values must be permitted.

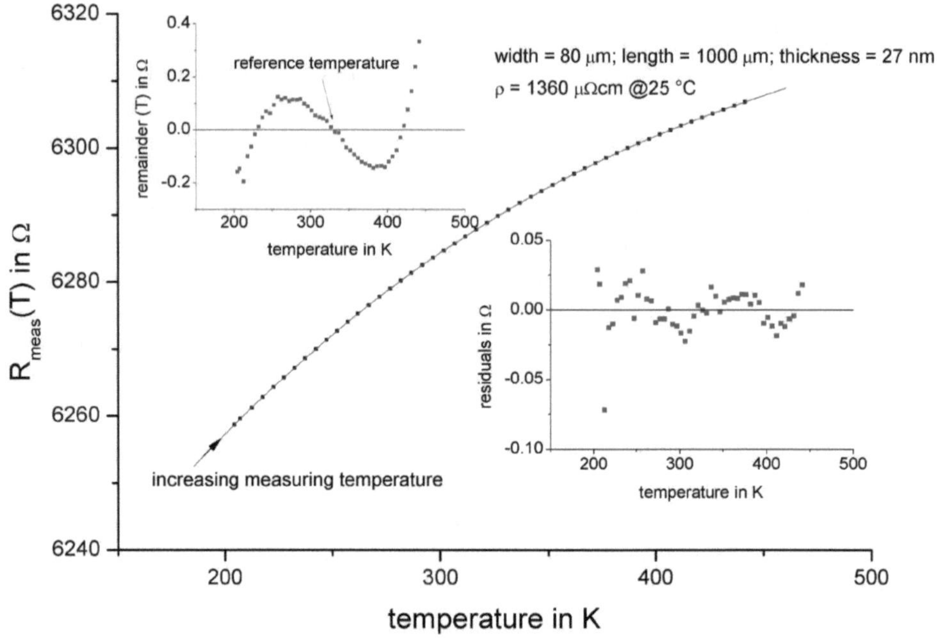

Figure 16.2: Example of a fit to experimental data.

Note: Fit to experimental data by means of Ansatz 16.1 without a remainder. The residuals of the purely quadratic fit are shown in the upper left corner in the small diagram whose ordinate is called remainder. Additionally using a series expansion with a reference temperature of 318 K according to Ansatz 16.2, the residuals shown in the diagram in the bottom right corner are obtained. Data provided with kind permission of Cicor Micro Electronics Reinhardt Microtech AG.

Ansatz 16.1 reflects that in real materials mechanisms are seemingly at work, which destroy translational invariance of the system leading to diffuse, temperature-dependent scattering of the electrons. Since three different temperature terms occur in the mathematical description, three physically different scattering mechanisms are taken into account by Ansatz 16.1.

The linear term might involve interactions with optical phonons at the measurement temperatures involved; the quadratic term might measure a spin-preserving scattering interaction, while the remainder term could possibly describe a mechanism, which is caused by the thermally induced disorder of the spin moments of the localized carriers. These are highly speculative conjectures. Nevertheless, if a_1 and a_2 were also allowed to assume negative values, Ansatz 16.1 always led to a satisfactory modeling of measured data.

Figure 16.2 shows a typical curve $R_{meas}(T)$ of a resistance measurement with increasing measurement temperature (initial curve) with curve fitting.[2] The remainder is

2 All resistance measurements as a function of temperature were first taken at increasing (initial curve) and then at decreasing temperature. Each sample was heated at 300 °C for 5 min prior to the first measurement.

shown in the additional diagram in the upper left corner and reflects the systematic deviations from the simple quadratic model. The remainder was expanded as a power series in the temperature difference $T-T_{remainder}$, where $T_{remainder}$ is a free parameter to be determined.

Many evaluations showed that in the series expansion two terms sufficed and that hence the heuristic

Ansatz 16.2 $$\text{remainder}(T) = a_3\,(T-T_{remainder})^3 + a_4\,(T-T_{remainder})^4$$

can be assumed. However, a mathematically stable determination of the reference temperature $T_{remainder}$ is difficult without an initial quadratic fit.

Measurements[3] according to "method 1" between –70 °C and +160 °C showed that the remainder was often zero around 320 K. Measuring[4] according to "method 2" between +20 °C and +200 °C systematically shifted the zero to a higher temperature, namely, to about 380 K.

It is noteworthy that the two experimentally observed reference temperatures are in the vicinity of temperature values listed in Table 5.3, which are obtained by dividing the reference temperature τ by the growth parameter g of the logistic model. Based on this observation, all experimental data in the temperature range from –70 °C to +160 °C were evaluated with a fixed $T_{remainder} = 318$ K ($\tau/3.57$), and all experimental data in the temperature range from +20 °C to +200 °C with a fixed $T_{remainder} = 378$ K ($\tau/3$). This defined setting never produced a convergence problem during the processing of all experimental data.

In Figure 16.2, in the lower right corner of the graph, residuals are plotted taking account of a remainder with a reference temperature $T_{remainder} = 318$ K. However, such a perfect fit, as can be seen in Figure 16.2, is not always achieved. Often "mavericks" are the rule at very specific temperatures. Such an anomaly is also visible in Figure 16.2 at ≈214 K.

3 Method 1:
Measurement in a large climate chamber (volume 190 l) at temperatures from −70 °C to +160 °C in steps of (5±1) °C. The temperature of the sample was in each case determined directly on the sample after the measurement of the last resistor. The temperature difference between the first and the last (usually the eighth) resisistance measurement was less than 0.2 °C.
The measurement of the resistance was carried out with measuring instrument X.

4 Method 2:
Measurement with a probe card on a temperature-stabilized (± 0.1 °C) brass plate at temperatures from 20 °C to 200 °C in increments of ±5 °C.
The measurement of the resistance was carried out with measuring instrument Y.

16.2.1 The ab plane: normalization of the measured data with the reference temperature τ

To compare the temperature dependence of resistors of variable geometries and compositions, the linear (a_1) and the quadratic coefficient (a_2) of Ansatz 16.1 were normalized with the temperature-independent coefficient (a_0) according to the following:

Definition 16.1 $a \equiv a_2/a_0$

Definition 16.2 $b \equiv a_1/a_0$

This standardization brings all measured data $R_{meas}(T)$ to a common scale, thereby facilitating comparison to detect a regularity or even a hidden law. The unit of the normalized parameter a is K^{-2}, and that of the parameter b is K^{-1}.

In Figure 16.3, data points that were calculated by means of Definition 16.1 or 16.2 are shown in an ab plane. By normalizing both axes, "universal curves" are seemingly obtained. For each data point, the error bars for 95% confidence interval were also calculated by linear error propagation. These error bars, however, become only clearly visible if a less expanded graphical range is chosen. The figure includes both data of increasing (red dots) and data of decreasing measurement temperatures (blue dots).

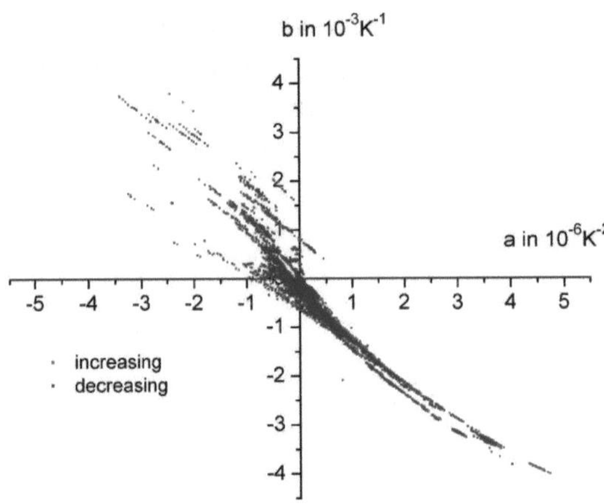

Figure 16.3: Entire measured ab plane.

Note: Graphical representation of the ab plane with the normalized quadratic term as the horizontal axis and the normalized linear term as the vertical axis. Each data point corresponds to a least-square fit to a measured temperature profile according to Ansatz 16.2. The measured resistors have both diverse geometries and variable compositions. Only the accuracy of the least-square adjustment was decisive, whether a curve fitting was transferred as a data point in the ab plane or not. Data provided with kind permission of Cicor Micro Electronics Reinhardt Microtech AG.

Only those data points were considered for which the coefficient of determination of the fit was better than 0.9999, the absolute error of the normalized quadratic coefficient (a) was less than 10^{-7} K^{-2} and the absolute error of the normalized linear coefficient (b) was less than 10^{-4} K^{-1}. In other words, only those temperature curves that were compatible with Ansatz 16.2 with a reference temperature $T_{remainder}$ of 318 K (chaos point τ/3.57) or 378 K (bifurcation point τ/3) are taken into account in the graphs.

It is striking that the data points in Figure 16.3 are not statistically distributed in the ab plane, but accumulate very densely in certain areas and do not occur at all in other areas. This may be related to the number of samples investigated or the temperature ranges passed through. However, it can also be a clue that certain areas really do not occur. Measurements of pure metals are missing in the survey, since no measurement data were available in this regard, because only poorly conductive materials were an issue in these investigations.

Figure 16.4 shows the same data as that in Figure 16.3 with related error bars and additional curves of linear functions b = f(a). These functions were added as an aid to the eye. They divide the ab plane for the point pairs (a, b) into different areas. In some cases, the measuring points pile up only on one side along a functional line. In other cases, they scatter around the functional line. These interpretational lines were not determined by the method of least squares, but arose heuristically by trial and error.

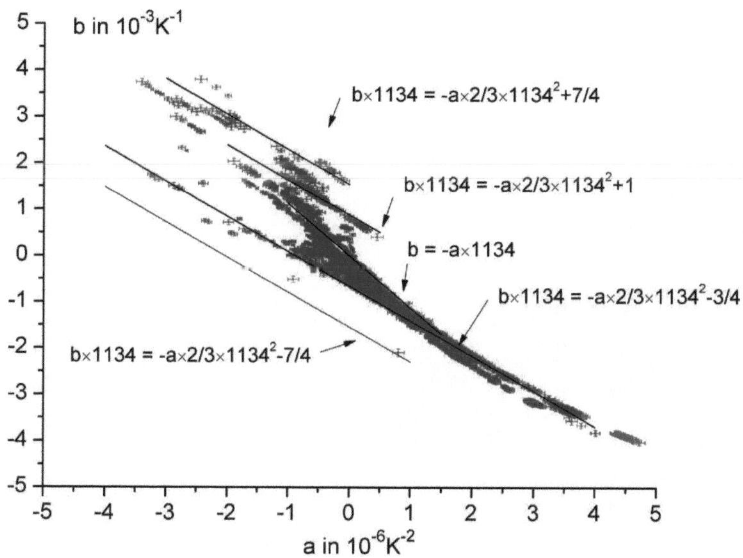

Figure 16.4: Limited data representation of the ab plane with error bars.

Note: Similar graphical presentation as Figure 16.3, but with error bars and with a different intersection of the coordinate axes. In addition, arbitrary interpretational lines for the eye are shown, which shall illustrate the meaning of the reference temperature τ. Data provided with kind permission of Cicor Micro Electronics Reinhardt Microtech AG.

It is noteworthy that only by using the reference temperature τ of ≈1,134 K and simple fractions p/q a model-setting description of the data points could be achieved.

Figure 16.5 is a partial enlargement of Figure 16.4. Here it becomes more apparent that certain functional lines subdivide the ab plane into regions where no or only very isolated data points can be found. For data points for which the normalized quadratic parameter a is zero, the temperature dependence is linear within the measured temperature range.

The most negative slope of all observed temperature curves with purely linear behavior amounts to $\approx -6.61 \cdot 10^{-4}$ K^{-1}, which is without sign equivalent to a reciprocal b of about (4/3) 1,134 K. Both the positive and the negative value of this characteristic point are shown as black squares in Figure 16.5. Assuming geometrized units according to Hypothesis 5.1, these characteristic points are equivalent to ≈± (3/4) L, or ≈9.5 μm, respectively.

The area of the ab plane that marks the transition from the semimetal to the insulator is represented in Figure 16.6. This region shows two separate data sets and contains many data points of temperature curves with large a-values, that is, temperature curves that are curved convexly. The surface resistances of such samples are extremely high and the remainder of Ansatz 16.2 plays a more important role in the least-square adjustment than in low-ohmic samples. Additionally, the accurate measurement of the effective temperature of the sample is extremely crucial here since the temperature coefficient is quite large.

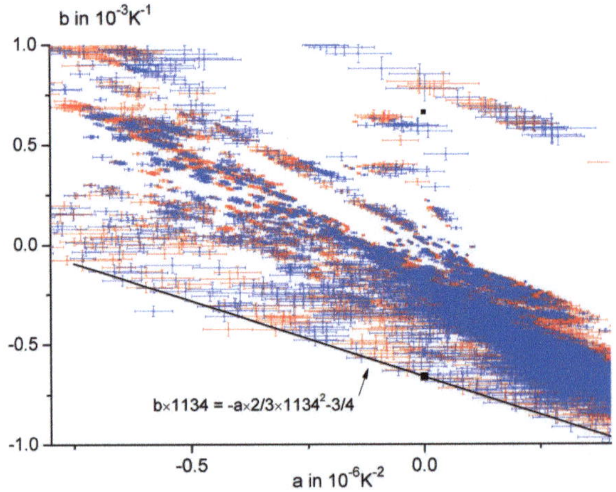

Figure 16.5: Enlargement of Figure 16.4.

Note: An enlargement of Figure 16.4 with an interpretational line, which both shows how densely packed the data points are in certain areas, as well as graphically sets out that the temperature (4/3) τ must have some physical meaning. Each cross line represents an evaluated resistance measurement. Data provided with kind permission of Cicor Micro Electronics Reinhardt Microtech AG.

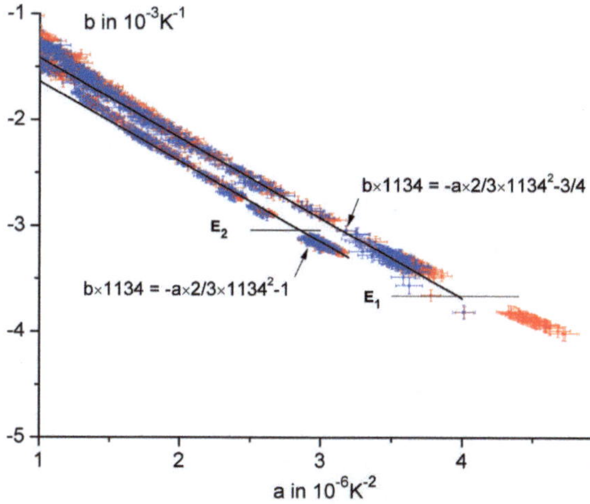

Figure 16.6: ab plane for measurement samples with "insulating behavior".

Note: Graphical representation of the relationship between linear and quadratic terms for "nonmetallic" behavior. The horizontal lines E_1 and E_2 plotted in the chart represent certain b-values and are explained in the text. Data provided with kind permission of Cicor Micro Electronics Reinhardt Microtech AG.

In Figure 16.6, two characteristic b-values are also drawn as horizontal bars E_1 and E_2. These b-values are explained in more detail in Table 16.1, with their corresponding reciprocal values. As a reminder, b-values are related to the linear part of the temperature profile. Is it a coincidence that there are no measured values in the environment of a reciprocal b-value of ≈273 K (E_1)? Why do all reciprocal b-values that are greater than ≈273 K relate almost exclusively to initial curves (red dot cluster)? Why are the measured values missing in the environment of the reciprocal b-value (E_2), which plays a role in the logistic model?

Table 16.1: Characteristic points.

	$-b$ (K^{-1})	$-b^{-1}$ (K)	$-b$ (in units of τ)
E_1	$\approx 3.66 \cdot 10^{-3}$	≈273	≈4.15
E_2	$\approx 3.04 \cdot 10^{-3}$	≈329	$\approx 3.45 \approx 1 + 6^{1/2}$

Note: Characteristic points of Figure 16.6.

The data points in Figure 16.2 represent typically metallic behavior with a concave temperature profile, analogous to the measurement curve shown in Figure 16.7. Unfortunately, there are fewer measurement data available in this region to make conclusive statements. In the figure there are also drawn interpretational lines for the eye to underpin the importance of fractional numbers. Here, too, it is observed that the measurement data are not a statistical cluster of points, but have a visible systematics.

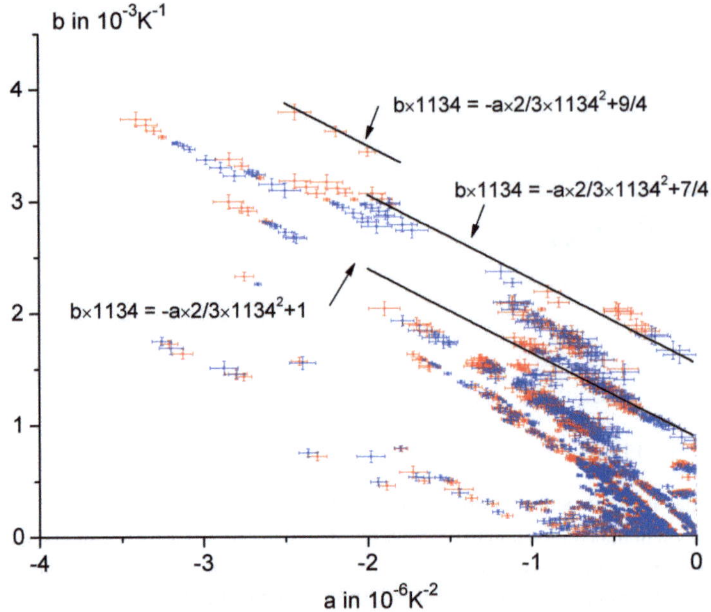

Figure 16.7: ab plane for measurement samples with "metallic behavior".

Note: Separate presentation of measuring points of the first quadrant of Figure 16.3 with interpretational lines. Note that the linear temperature coefficient b of the metals aluminum, gold, copper or platinum amounts to $\approx 4 \cdot 10^{-3}$ K^{-1}, yielding an axis intercept at a = 0 of approximately $4 \cdot 10^{-3} \cdot 1134 \approx 9/2$. Data provided with kind permission of Cicor Micro Electronics Reinhardt Microtech AG.

16.3 Geometry dependence of resistivity

For a dense material, it is phenomenologically proven that the specific resistance (ρ_{el}) usually depends only on the temperature and the composition, and exhibits no dependence on the extension of the sample. In other words, according to Ohm's law, it can be assumed that the measured resistance value halves when the geometric width of the resistor is doubled. However, if a characteristic path length of the conduction electrons is larger or comparable to a sample dimension, the resistivity may become geometry dependent and Ohm's law loses its validity.

What is well known is that the specific resistance of a film consisting of a few atomic layers increases compared to dense samples. The model of Fuchs [16.8] and Sondheimer [16.9] analytically explains with quasiclassical arguments why artificially limiting the mean free path can cause an increase in resistivity over the value of the massive sample. In semiconductor industry [16.10], it is known that the electrical properties of copper change and the size effect [16.11] becomes more and more relevant when copper is used for nanostructures. And, nanocrystalline metals or alloys usually exhibit an increase in resistance with smaller crystals.

Interesting experiments showing geometry-dependent resistivities were reported by Benjamin et al. [16.12] on discontinuous gold films on glass near the metal-insulator transition. On approaching the metal-insulator transition, they obtained a reduced resistivity for short, wide geometries compared to macroscopic dimensions. Conversely, they predicted that increased resistivities will be observed in long, narrow geometries, and pointed out that similar phenomena must occur in other systems as well on approaching the metal-insulator transition.

If the composition or the film thickness is systematically modified while maintaining the lateral geometry, it is possible to determine whether the classical resistance formula based on simple geometrical considerations is actually valid. For a constant thickness and laterally structured resistors, the phenomenological relationship according to

Ansatz 16.3 $$R_{meas} = \rho_{el}\, l\, (wd)^{-1} = R_{sq}\, l\, w^{-1}$$

holds for macroscopic systems. Such resistors have a uniform thickness (d) and, in a first approximation, depend only on the ratio of the length (l) to the width (w). Ansatz 16.3 is not a physical law that can be deduced from physical principles, but follows from the experience of macroscopic systems.

The sheet resistance R_{sq} is a very practical quantity, which strongly ($\sim d^{-1}$) depends on the thickness (d), but usually does not show a pronounced relation to lateral dimensions. However, the scale invariance of R_{sq} (d) to length changes is not always given and it must be assumed that only a generalized dependence of R_{sq} (d,l) can correctly describe experimental data of limited systems if relevant length scales are on the order of magnitude of the dimensions of the sample.

16.3.1 Width dependence at constant length and constant temperature: the existence of critical widths

To test Ansatz 16.3, several resistors of constant length were fabricated on the same substrate to examine the width dependency. For all widths, the length was ≈1 mm. Since all resistors are located within a very small area of the substrate, it can be assumed that both the material composition and the film thickness of all resistors are nearly identical. The material composition of the different samples was such that a range of sheet resistances between 70 Ω (210 μΩ cm at 30 nm)[5] and 60 kΩ (180 mΩ cm at 30 nm) was covered at film thicknesses of the samples of about 30 nm.

5 Materials with resistivities greater than approximately 100 μΩ cm are no longer considered as conductors, but as semiconductors.

Each resistance value was determined at ≈20 °C as an average of the measurement at increasing and at decreasing temperatures. Only those measured values were considered whose difference in the resistance value caused by temperature hysteresis or measurement error was less than 0.3%. From measured resistance values (R_{meas}) for each width, the corresponding sheet resistance R_{sq} can be calculated by means of Ansatz 16.3, which should actually be equal for all widths according to Ohm's law. Experimentally, however, large fluctuations in the surface resistances were surprisingly obsereved, depending on which width was used for the calculation of R_{sq}.

To compare samples with different surface resistances, all measured resistance values of the different widths were normalized with the surface resistance of the largest width (100 µm), which corresponds to a tenth of the length held constant for all eight widths. In this way, all information about the individual system is lost, but possibly some universal geometrical attributes can be identified as a statistical approach.

In Figure 16.8, the experimental situation of several resistors with *w* equal to 10, 20, 30, 50 and 80 µm is shown graphically by frequency distributions with a column width of 5%. This representation impressively reflects that the normalized values of the surface resistances considerably fluctuate, and that the ohmic Ansatz 16.3 adopted is not correct because not all values are equal to 1. The significant deviations from 1 are not due to experimental errors, since the lithographic accuracy was ±1 µm. Rather, different mean free paths of the conduction electrons must seemingly be attributed to it. The classical description of charge diffusion apparently collapses, as additional localization and interaction effects of bound states become important. A more suitable scaling between sheet resistance and width than the analytical Ansatz 16.3 is difficult to extract from the data. It seems that the surface resistance, and with it the fluctuation margin, systematically increases with smaller widths compared to macroscopic dimensions.

The collapse of Ohm's law implies that the electrical properties of resistors of widths on the order of micrometers are difficult to predict. Resistive geometries of the dimensions of a few micrometers undoubtedly suggest that even on the micrometer scale the classical formula for describing the geometry dependency can no longer be used, and requires a rethinking and technical precautions. The most surprising thing is that there obviously exist long-range electron scattering mechanisms whose length scales are in the micrometer range.

From Figure 16.8, it is difficult to gather anything in common or even to find a mathematical law about the systems studied. The finding that the sheet resistance tends to increase with decreasing width suggests, analogous to critical phenomena, to try as a functional dependency at constant length and constant thickness a simple scaling law for $R_{meas}(w)$ according to the following:

Ansatz 16.4 $$R_{meas}^{-1} = konst\,(w - w_c)$$

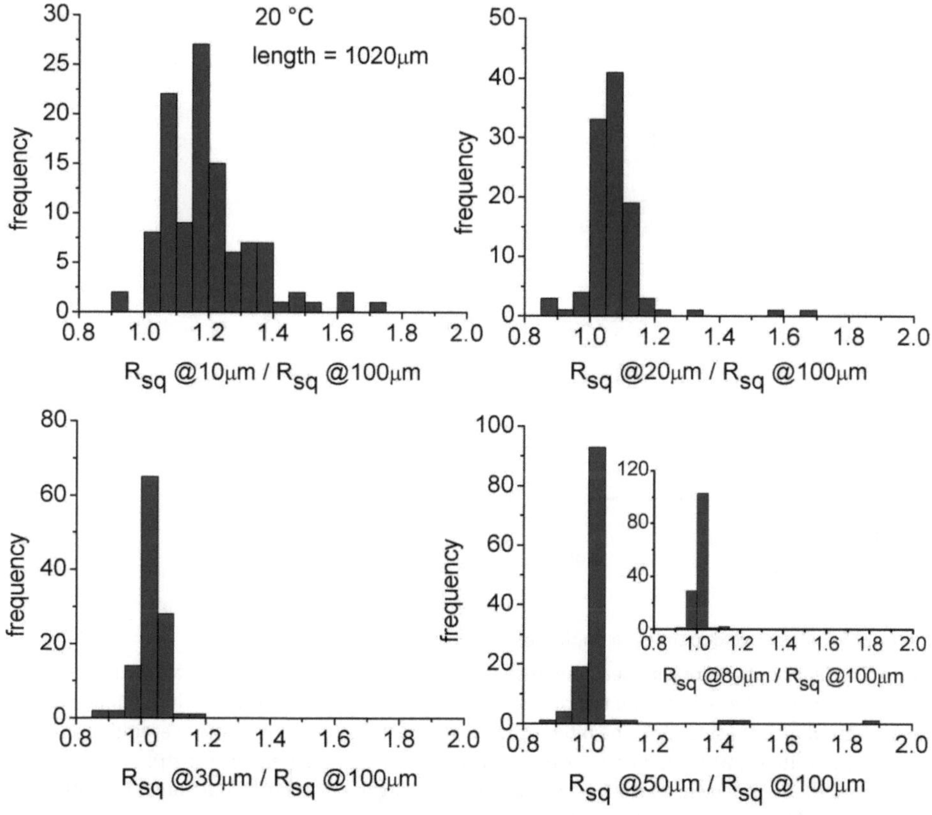

Figure 16.8: Breakdown of Ohm's law.

Note: Representation of the frequency distributions of normalized surface resistances with the resistance width as a parameter. Each surface resistance is a multiple of the surface resistance of the resistor with the largest width (100 μm) of the same sample. It can clearly be seen that with decreasing width the sheet resistance shows more deviation from the theoretical value of 1 and tends to increase. The breakdown of Ohm's law is visually visible. Data provided with kind permission of Cicor Micro Electronics Reinhardt Microtech AG.

Using Ansatz 16.4, the measured resistance is determined by three macroscopic quantities, such as the critical exponent, which is set equal to –1 on the basis of experimental data; the correlation width w_c as a critical dimension and the constant *konst*, which depends on the material, the measurement temperature and the thickness and the length of the resistor. Both the characteristic width w_c and the constant *konst* can be determined by linear regression. Such a regression with an extrapolation of R_{meas}^{-1} to zero is shown in Figure 16.9.

According to Ansatz 16.4, various samples of different compositions were evaluated and it was found that the model was very appropriate to describe the width

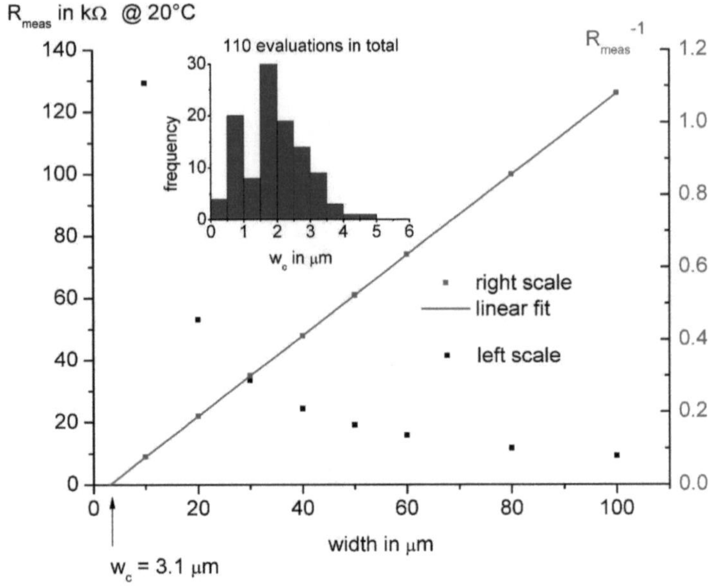

Figure 16.9: Resistive dependence on the resistor width.

Note: The black dots represent measured resistance values of a single sample as a function of the resistance width. The red dots are the corresponding reciprocal values through which a linear fit was made according to Ansatz 16.4. The intersection of the best-fit straight line with the width axis gives the critical width w_c, which is marked with an arrow. As an inset, the frequency distribution of all critical widths w_c is shown, which were evaluated according to Ansatz 16.4. Data provided with kind permission of Cicor Micro Electronics Reinhardt Microtech AG.

dependency, if certain "anomalies" that were occasionally seen at certain widths were not taken into account. A reciprocal value is called an "anomaly" if it deviates from the linearity of Ansatz 16.4. The "anomalies" are not due to manufacturing errors since an optical inspection of the samples with a microscope clearly showed no lithographic defects. In a second measurement, mostly (but not always) the same effect could be recognized. Only subsequent heat treatments made the "anomalies" to disappear again, that is, the reciprocal resistive value again followed a linear behavior according to Ansatz 16.4 similar to other measuring data points. A linear regression with an anomaly is shown in Figure 16.10. It appears as if an "anomaly" is part of another transport mechanism that is observed only for a specific geometry and not for all the others. A dimensional change of the system resulted in a change of the behavior.

The number of data points per adjustment was always at least six out of a maximum of eight resistance values. The coefficient of determination, which measures the quality of the linear regression, was always greater than 0.9999. Smaller coefficients of determination, which sometimes occurred because Ansatz 16.4 failed, were

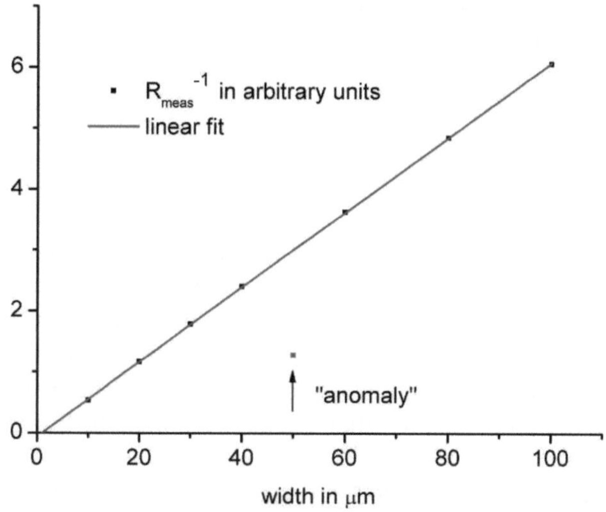

Figure 16.10: Breakdown of the scaling law.

Note: Illustration of the fact that a reciprocal resistive value (in arbitrary units) of a certain width lies not on the best-fit straight line, which otherwise describes the reciprocal resistive values of the other widths. Obviously, at 50 μm the scaling law breaks down, which otherwise allows good modeling of all other widths. Data provided with kind permission of Cicor Micro Electronics Reinhardt Microtech AG.

not included in the data collection of w_c. The frequency distribution of w_c of several samples at a measurement temperature of ≈20 °C is inserted as a smaller diagram in Figure 16.9. The choice of 20 °C as a reference temperature is arbitrary. The correlation widths w_c are temperature dependent, but their frequency distribution does not change significantly when other measurement temperatures are taken into account. Above all, the upper limit of w_c is not affected by other temperature choices. The smallest surface resistance of all the samples, calculated by means of the resistance of the largest width (100 μm), is ~ 74 Ω; the largest sheet resistance amounts to ≈$6.6 \cdot 10^4$ Ω. The material composition thus covers a range from the semimetal to the semiconductor. Highly conductive metals were not considered in the investigation, because they exhibit very small correlation widths w_c.

The results described make it clear that all correlation widths w_c evaluated with Ansatz 16.4 were always smaller than ≈4.5 μm. It seems that the width of ≈4.5 μm is the limit length transversely to the current direction because this value was never exceeded in all evaluations. The modeling of width dependences at temperatures other than the reference temperature of 20 °C was no exception in this regard. Although the correlation widths are temperature dependent, they never exceeded a certain limit. Does this outcome indicate a universal principle? Is this limit length possibly determined by the characteristic length Λ of ≈4.33 μm described in Section 5.2? Is it the maximum width of the conducting channel for current flow? Or is it just an evaluation artifact of too few case studies?

16.3.2 Length dependence at constant width and constant temperature: the existence of critical lengths

In analogy to Ansatz 16.4, we use

Ansatz 16.5 $R_{meas} = konst\,(l - l_c)$

for the length dependence of $R_{meas}(l)$ at constant width and constant thickness. Similar to the data used for evaluating the width dependence, the film thicknesses from sample to sample vary between 20 and 50 nm. However, each individual sample has a homogeneous layer thickness distribution on the measuring area. With Ansatz 16.5 the measured value $R_{meas}(l)$ is approximated by a simple scaling law, but with a critical exponent of 1 and a correlation length l_c that can be determined by linear regression.

As a result of the data evaluation with Ansatz 16.5, no significant differences have been found as compared to the use of Ansatz 16.4. A typical data evaluation is shown in Figure 16.11. The linear regression was always performed with eight resistance values measured on the same substrate, as in no case a resistance value due to "anomalous behavior", that is, due to the deviation from linearity, had to be deleted. For

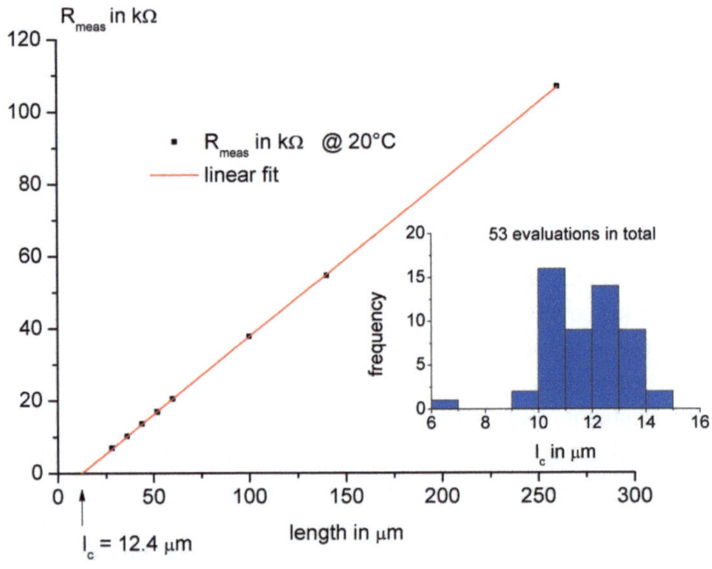

Figure 16.11: Resistive dependence on the resistor length.

Note: Graphical representation of the evaluation of the length dependence of a single sample at constant width. The smaller diagram shows the frequency distribution of the critical length l_c, which was obtained by linear regression to eight resistance values on the same sample. Data provided with kind permission of Cicor Micro Electronics Reinhardt Microtech AG.

the calculation of l_c only adjustments that had a coefficient of determination greater than 0.9999 were used.

The frequency distribution of the fitting parameter l_c of all evaluated samples of different material composition is also incorporated in Figure 16.11. The smallest sheet resistance of the samples calculated by means of the resistance of the longest length (260 µm) is ≈335 Ω; the largest sheet resistance amounts to ≈$3.5 \cdot 10^6$ Ω. As in the case of the width dependence, highly conductive metals were not considered since only semimetals and semiconductors were of interest.

The correlation length l_c also seems to be limited by an upper limit length as in the case of the width dependence. Data analysis revealed a value of ≈15 µm for this limit. This value can be experimentally flawed, since the contacts were set back by approximately 5 µm compared to the width-restricted resistive layer due to lithographic reasons. This situation can be clearly seen in Figure 16.32, which consists largely of three different shades of gray. The brightest gray tone, recognizable on the left and on the right edge of the picture, symbolizes the displaced contacts of the leads to the resistor. What influence this displacement has on the correlation lengths is not known. The length of the resistor was defined as the distance between the two contacts, although the constriction of the resistive layer did not occur over the entire resistor length. Despite this uncertainty and the far too few case studies, it can be assumed that the limit length of ≈15 µm in the current direction could be determined by the characteristic length L of ≈12.7 µm described in Section 5.2.

16.3.3 Temperature behavior with variable geometry: pronounced size effects

In some measuring runs, it has been observed that the temperature dependence of the resistance $R_{meas}(T)$ can significantly vary with width if the film thickness and the length are kept constant. The change in the temperature behavior was not due to a modification of the composition, but was solely caused by a modified geometric boundary condition, since the resistors compared came from the same sample. They must, as discussed in previous sections, have approximately the same composition, because all the resistors compared were located within an area of 4 mm × 0.2 mm or smaller during manufacturing. The observations described in this section are closely linked to the anomalies discussed in Section 16.3.1 (Figure 16.10). In any case, they reflect the influence of geometry on the measurement result in an even more concise manner.

In Figures 16.12 and 16.13 examples are given that show the temperature response of two resistors with a width of 15 or 20 µm, respectively. Both widths show a completely different temperature dependence, although a difference in width should not show such a pronounced difference in behavior according to classical understanding. In both cases, the narrower resistor behaved more like a semiconductor.

Comparing the 20 µm wide resistor in both representations reveals that in Figure 16.13 the resistor behaves like a typical metal with a linear temperature characteristic,

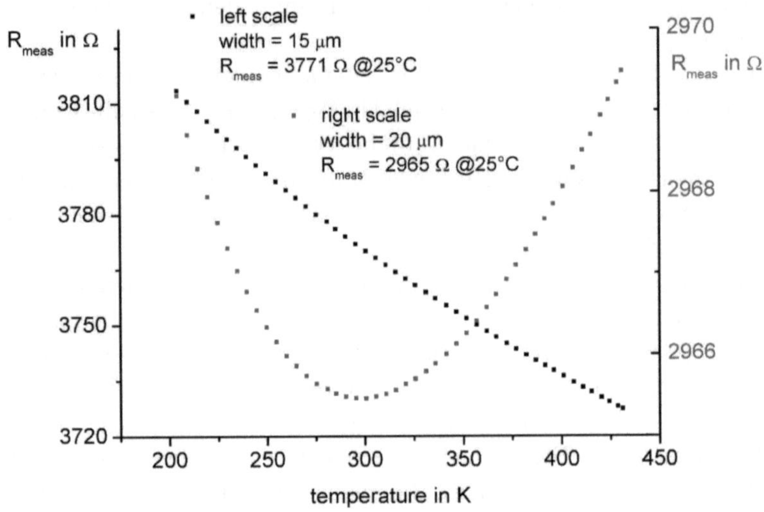

Figure 16.12: Resistive minimum at room temperature.

Note: This illustration is a model example for what effect the resistor width can exert on $R_{meas}(T)$, if the length (≈120 μm), the thickness (≈27 nm) and the composition are kept constant. Striking is the pronounced resistive minimum at room temperature, which is caused by a small change in the resistor width alone. The resistivity did not change significantly. Data provided with kind permission of Cicor Micro Electronics Reinhardt Microtech AG.

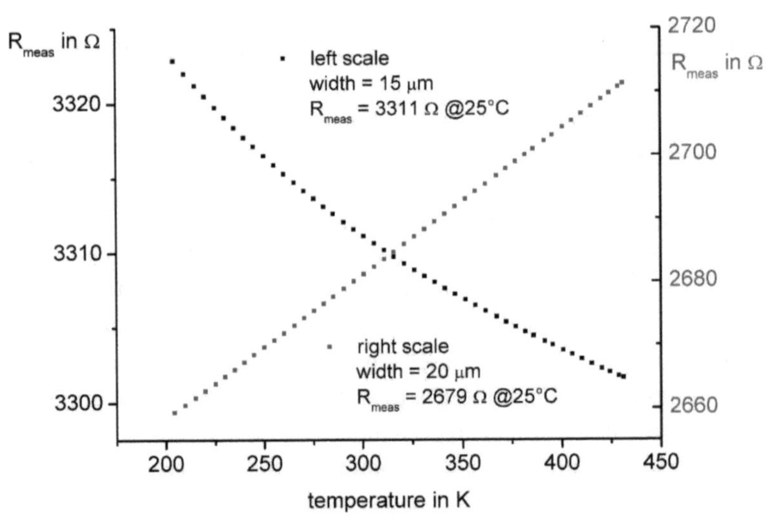

Figure 16.13: Size effect causes a change of sign of the TCR.

Note: Reversal of the sign of the temperature coefficient by the resistor width. A slight change in width switches from "semiconducting behavior" to "metallic behavior" with a linear temperature characteristic, while approximately maintaining the resistivity. Data provided with kind permission of Cicor Micro Electronics Reinhardt Microtech AG.

while in Figure 16.12 the resistor shows both metallic and semimetallic behavior within the measured temperature range. This behavior results in a characteristic resistance minimum similar to the resistance minimum of the "Kondo effect", which is, however, not at 30 K (Cu), but at 300 K.

A small difference in width is obviously capable of producing a significant change in the functional form of the temperature dependence while maintaining the length (≈120 μm), the thickness (≈27 nm) and the composition. In other words, the transport law seemingly changes if the resistor is scaled from 15 to 20 μm in width, and thus the charge carriers are limited by a slightly different cross section.

The width dependence between 15 and 20 μm of the systems shown must mark the proximity of a critical point with the width as a control parameter, where the metallic term gets ever more relevant in describing the temperature dependence, and therefore a radical change of the temperature behavior of the system can take place. In both cases, in contrast to the temperature dependence, the resistivity of ≈1,250 μΩ cm at 25 °C did not change significantly. That is, all measured resistance values on the sample scaled approximately according to Ohm's law.

In Figure 16.14, a system is graphically set forth where both the resistivity and the temperature behavior varied widely with a small change in width. Although the nearly perfect parabolic shape of the temperature profile was maintained for both widths, the resistance maximum was significantly shifted by the width modification. By halving the width, the resistance of the sample did not increase by a factor of two, as might be expected according to Ohm's law, but by a factor of four.

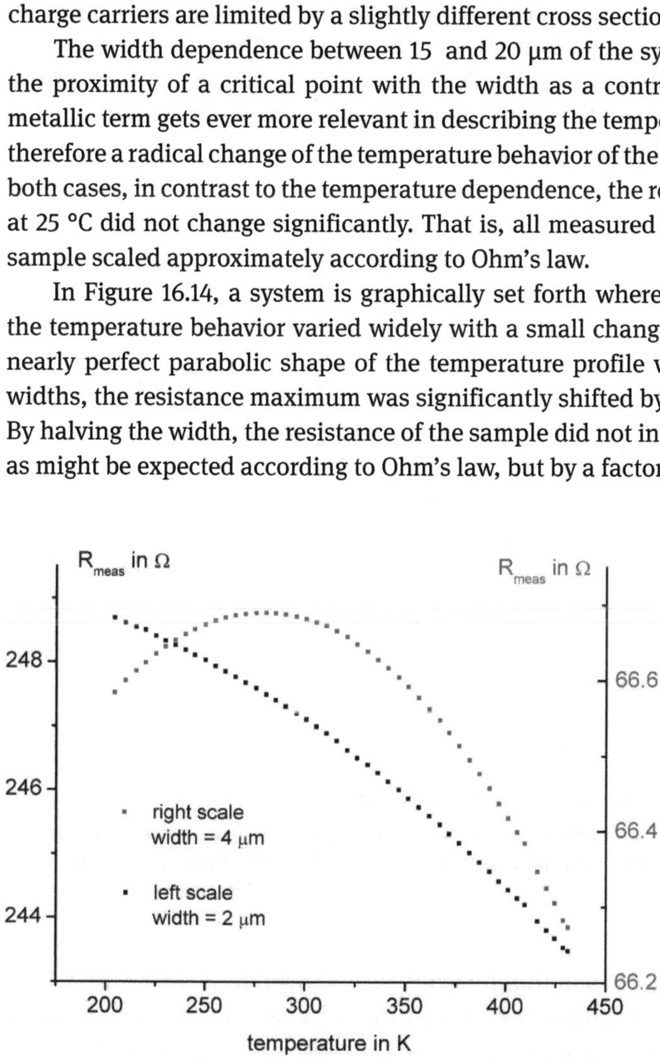

Figure 16.14: Lateral size effect causes a change in sheet resistance.

Note: Violation of Ohm's law with a change in the resistor width. Note that the curves refer to different scales. Data provided with kind permission of Cicor Micro Electronics Reinhardt Microtech AG.

A sensitive changeover to a modified transport behavior with the width as a control parameter is also displayed in Figure 16.15. At half the width, the resistance of this system increased by approximately a factor of four, which corresponds to twice the value expected on the basis of Ohm's law. The temperature coefficient changed sign while maintaining the curve shape opened downward. At the same time, the temperature profile of the nonmetallic phase that belongs to the thinner resistor with a negative TCR shows a "room temperature anomaly" observed in other systems as well, but which is completely absent in the metallic phase. It seems systems behave asymmetrically in this respect, because the "room temperature anomaly" has never been observed in a pronounced form in metallic systems. The example shown in Figure 16.15 demonstrates again, as already discussed in Section 16.3.1, that resistive properties depend not only on the composition, but also – in a complex way – on the geometry of the resistive layer.

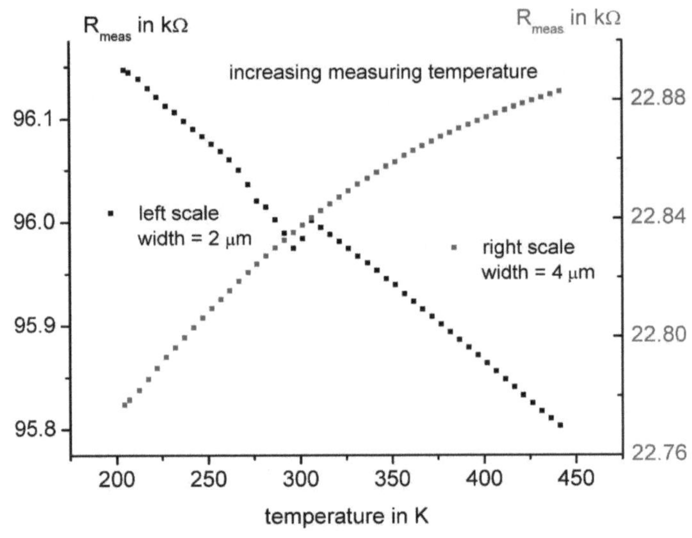

Figure 16.15: Sign change of the TCR and variation of the resistivity.

Note: A change in width causes both a clear sign change of the temperature coefficient and a variation of the resistivity. Data provided with kind permission of Cicor Micro Electronics Reinhardt Microtech AG.

Size-induced phase transitions in thin films are well known in the literature [16.13]. In contrast to the transitions described previously, however, they are mostly based on changes of local nanosized material clusters. In any case, between the micrometer and the atomic scale lies a fascinating and important field of physics and chemistry of condensed matter. It is well known that metal particles with diameters of less than 10 nm have fundamentally different chemical and physical properties than macroscopic parts. For example, the conductivity of interconnects becomes size dependent if dimensions are in the nanoscale. However, observing size effects in the micrometer range has provided surprising results.

Phase transitions in materials are caused by interactions to which characteristic lengths are attributed. The examples previously shown imply that there must be characteristic lengths not only in the nanometer range, but also in the micrometer range, because sample geometries of the order of microns can drastically influence electrical transport properties. The question is whether characteristic lengths in the micrometer range for the charge carriers prevail in all matter, and are therefore universal, or whether they only apply to a certain class of materials. In any event, it is astonishing that electrical transport properties exhibit marked changes by a different confinement, that is, by tailoring simple, rectangular geometries in the micrometer range without altering the chemistry of the amorphous systems.

16.4 Thermal hysteresis as a result of nonlinear behavior

Many measurements showed that the curves at increasing and at decreasing temperatures varied, which was often clearly visible to the naked eye, and the resistance value depended on the whole measuring history of the sample as with the magnetization. Such a behavior, which was more or less pronounced after each heat treatment, indicates that the systems are not in thermal equilibrium. Obviously, different interactions and feedbacks arise with an increase (heating process) or a decrease (cooling process) of the temperature. The consequence is nonlinear behavior with magnetic order-disorder transformations.

By a heat treatment and a subsequently rapid cooling of the sample, a complicated energy landscape is formed, which disappears only after passing through many different temperature cycles. In the end, the system asymptotically corresponds to a correlated (entangled) state with the energetically most favorable distribution of matter whose atomic arrangement deviates from statistical chaos. Since the resistive layers can be considered homogeneous by microscopic examination, the observed thermal hysteresis behavior is not due to a precipitation process of a new phase. Repeated thermal load (thermal processing) causes the hysteresis loop to gradually narrow in most systems. It cannot be completely eliminated in those systems that are incapable of forming a deepest state of energy.

The remainder according to Ansatz 16.2 was always the most pronounced in the initial curve (heating process) and healed or at least became very small with the number of temperature cycles undergone. This temperature-dependent parameter probably has something to do with the concepts of order and disorder (entropy) and measures the amount of chaos that exists in the system. The greater the disorder, that is, the higher the entropy, in the system, the more significant the contribution of the remainder becomes in the measurement of the temperature dependence.

This implies that the reference temperature, which was introduced as a free parameter in the remainder defined by Ansatz 16.2, also concerns the same topic. In the usual way, disorder or chance is only allowed at the microlevel, and macroscopic systems are described by deterministic laws. Is the reference temperature of Ansatz 16.2 a link to the microcosm? Does Ansatz 16.2 allow the description of macroscopic samples as powers of temperature, since power functions of the form $A \cdot T^B$ are scale invariant?

In Figure 16.16, aftereffects of temperature cycles are graphically shown: The left half shows the first cycle, and the right half of the graph shows the third cycle for increasing and decreasing measuring temperatures. It can be seen from the plots of the residuals inserted in Figure 16.16 that the description of the temperature dependence within the measuring temperature range is better for the third cycle than for the initial cycle, and the errors are statistically distributed after thermal cycling.

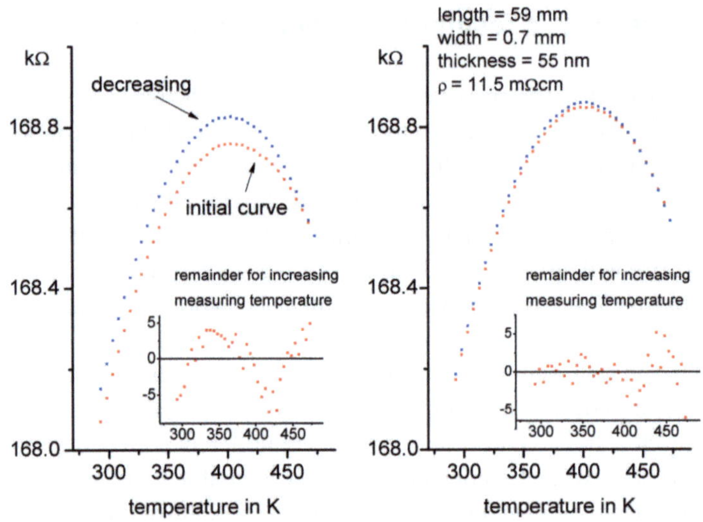

Figure 16.16: Thermal hysteresis.

Note: The difference between the resistance measurement curve with increasing measurement temperature (initial curve) and the resistance measurement curve with decreasing measurement temperature is called thermal hysteresis. It depends on the start and end temperature and the number of measuring cycles within the temperature range. Thus, it is a measure of the thermal "balance" or order of the sample. Data provided with kind permission of Cicor Micro Electronics Reinhardt Microtech AG.

The description of the temperature profile by a remainder consisting of two power functions was not successful in all measurements. Sometimes Ansatz 16.2 failed with such a simple analytic function. Figure 16.17 shows two diagrams where the typical thermal hysteresis behavior seen in Figure 16.16 occurred only in certain temperature

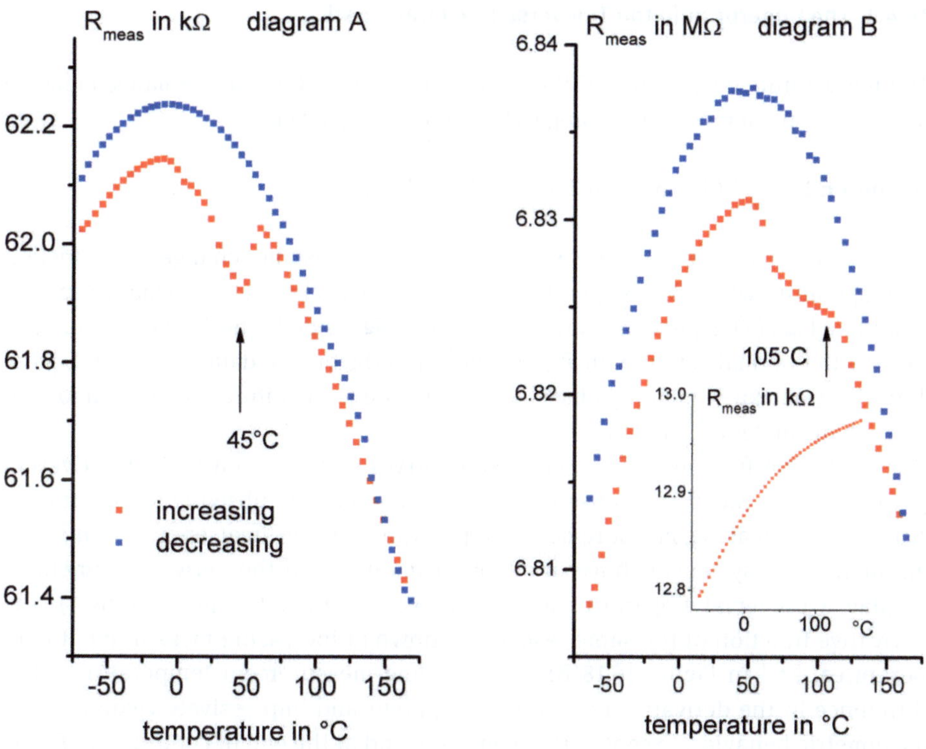

Figure 16.17: Anomalies in the thermal hysteresis.

Note: The hysteresis-inducing heat treatment is often accompanied by anomalies that cannot be described by a simple remainder according to Ansatz 16.2. The hysteresis effect does not occur "uniformly" at all measuring temperatures, but is noticeably different at certain measuring temperatures. (A) and (B) show two curves of different samples, where this situation is clearly visible. The anomalies are also strongly dependent on the geometry of the resistor. In (B), a small graph is inserted as a comparison to the curve of the long resistor, showing an initial curve of a much shorter resistor on the same sample. Data provided with kind permission of Cicor Micro Electronics Reinhardt Microtech AG.

ranges and markedly deviated therefrom at temperatures between ≈0 °C and ≈100 °C. Is this experimentally observed deviation from Ansatz 16.2 a fingerprint of the behavior of Formula 5.2, whose properties are given in Table 5.3? The temperatures 45 °C or 105 °C at least seem to play a role.

In Figure 16.17 it is also shown that at constant width long resistors are more affected by the anomaly than short resistors. In Figure 16.17(B), the initial temperature curve of an adjacent, approximately factor of a thousand shorter, resistor of equal width was added as a comparison, which had no anomaly. The resistors providing the two temperature curves shown in Figure 16.17(B) were on the same sample.

16.4.1 The fingerprint in the TCR versus T characteristic

The temperature coefficient (TCR) of a resistor is defined as the normalized change $\delta R_{meas}/R_{meas}$ per temperature change δT in units of K^{-1}, given by

Definition 16.3 $TCR = \delta R_{meas}/\delta T/R_{meas} = R'_{meas}/R_{meas}$

Sensitive TCR measurements are well suited to study resistance changes as a function of temperature when the temperature difference δT is chosen to be small. A purely quadratic $R_{meas}(T)$ dependence gives almost a straight line in the TCR(T) diagram, as long as the normalization with $R_{meas}(T)$ does not distort the data too much. Such a behavior was experimentally observed over a wide temperature range in almost all high-ohmic materials studied.
The derivation functions (R'_{meas}) of initial curves often show a typical derivation pattern, which was mostly absent or less pronounced in subsequent temperature curves at decreasing measurement temperatures. When heat energy is initially applied, many systems behave in a similar manner and the derivative functions display a universal fingerprint for the nonequilibrium. In Figure 16.18(A), the derivative function of the same resistor is shown at increasing measurement temperatures, and in Figure 16.18(B) at decreasing measurement temperatures. The difference in the derivative functions is apparent and impressively visualizes the asymmetric behavior. Except at the beginning and at the end of cooling, the TCR is very linear.[6]

A significant TCR deviation from linearity during the initial heating cycle was also observed in many derivative curves of other samples. Remarkable is the fact, shown in Figure 16.18(C) and (D), that resistors with TCR<0 and resistors with TCR>0 with respect to ≈45 °C, that is, the chaos point $\tau/3.57$ (Table 5.3), reveal nearly mirror-image derivative functions. This indicates that probably the measurement system is not the cause of this experimental finding, and so another reason must be assumed.

The major change in the temperature behavior at ≈−55 °C, which is shown Figure 16.18, was observed in most data evaluations. Although this change in the temperature behavior is at the boundary of the measuring temperature range, it may be a hint for a characteristic transition. However, a simple correlation to the reference temperature τ could not be found.

---

6 For data analysis, the smoothing procedure of Savitzki–Golay was used. This algorithm smooths the measured data by a local least-square adjustment to a polynomial function on the measured value and measured values to the right and left of it. A second-degree polynomial with two neighboring points each was used and the first derivative at the measured value was calculated from the coefficients of this polynomial.

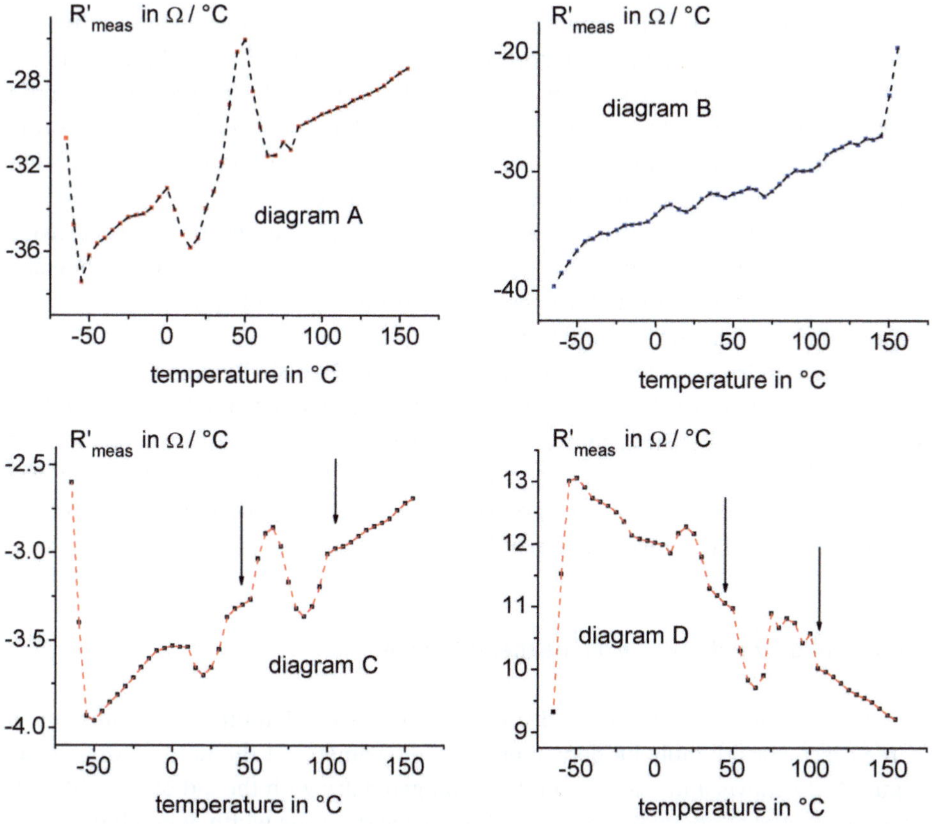

Figure 16.18: The fingerprint in the derivative function.

Note: For many resistance measurements, a fingerprint displayed in (A), (C) and (D) was observed in the derivative function of the initial curves. This dactylogram in the initial curve was detected in both semiconducting and metallic resistors. (C) and (D) show that curves with TCR > 0 or TCR < 0 can almost be converted into one another by flipping up and down at the chaos point at ≈45 °C. The vertical arrows mark 45 °C (τ/3.57) or 105 °C (τ/3), respectively. (B) shows that the fingerprint of (A) heals up in subsequent temperature cycles. Data provided with kind permission of Cicor Micro Electronics Reinhardt Microtech AG.

In this context, the anomalous resistive behavior of chromium at room temperature shall be briefly addressed. Perhaps this phenomenon is related to the room temperature anomaly observed as shown in Figure 16.17.

In 1933, P. W. Brigman found an anomalous temperature dependence near room temperature in the resistivity of chromium, which assumes a minimum at ≈12 °C that shifts to higher temperatures with an increase in pressure. This anomaly was confirmed by H. Söchtig [16.14] by investigating chrome samples of different purity. Depending on the purity of the sample, the minima ranged from about –3 °C to 40 °C. He found that the thermal power, the thermal conductivity, the thermal expansion and the compressibility are also abnormal in the room temperature range. A structural phase transformation could be disproved by x-ray studies. H. Söchtig thought that the anomaly must be based on a process that changes the elastic forces between the atoms or the binding of the conduction electrons with no structural change occurring. Chromium is the only metal that has a negative electrical temperature coefficient within a narrow temperature range at room temperature.

Arajs et al [16.15] also observed a considerable thermal hysteresis in the electrical resistivity of chromium single crystals in the room temperature region. They suggested that the effect results form variations in the degree of spin ordering, which depend on both temperature and time.

Stebler [16.16] even found a discontinuity in the resistivity curve of chromium single crystals when the temperature was slightly changed near the Néel temperature. However, he found, using neutron-scattering experiments, no direct relationship between the resistivity anomaly and the magnetic order. The anomaly also showed considerable hysteresis behavior. The exact nature of the experimental findings could not be revealed by this investigation either.

16.4.2 Influence of the size on the thermal hysteresis

The thermal hysteresis in resistivity measurements of high-ohmic samples can in some cases considerably be dependent on the geometry of the patterned resistors. Figure 16.19 shows a marked dependence on geometry with the aid of four different insets of curves of unequal widths developing a very strong width dependency.
The thermal hysteresis in the initial curves disappeared noticeably at a width of ≈12 μm and was present again at both smaller and larger widths. It is perhaps no surprise that the most pronounced hysteretic effect occurred at a temperature of ≈45 °C. All resistors were juxtaposed, arranged widthwise on the same substrate and therefore have nearly the same composition and thickness. The sample, whose resistive behavior is outlined in Figure 16.19, is near the semiconductor-insulator transition, that is, just before the electrons are fully localized. However, complete localization or disorder cannot yet be assumed, since the sheet resistance, although very high, is still $\approx 2.5 \cdot 10^6\ \Omega$.

Another example of pronounced geometric effects is shown in Figure 16.20. For three resistors, the resistance values changed in a hysteretic manner during a temperature cycle, while no pronounced thermal hysteresis could be detected for a resistor of width ≈13 μm. In addition, there are large differences in resistance values among the resistors, which are incompatible with Ohm's empirical law that is otherwise always valid for macroscopic samples. The resistors with a marked thermal hysteresis have a temperature curve shape opened downward during the cooling phase, while the resistor with a very small hysteresis effect has a temperature curve shape opened weakly upward. All resistors are part of the same sample, that is, they all have about the same thickness, the same composition and differ only in their lateral extension.

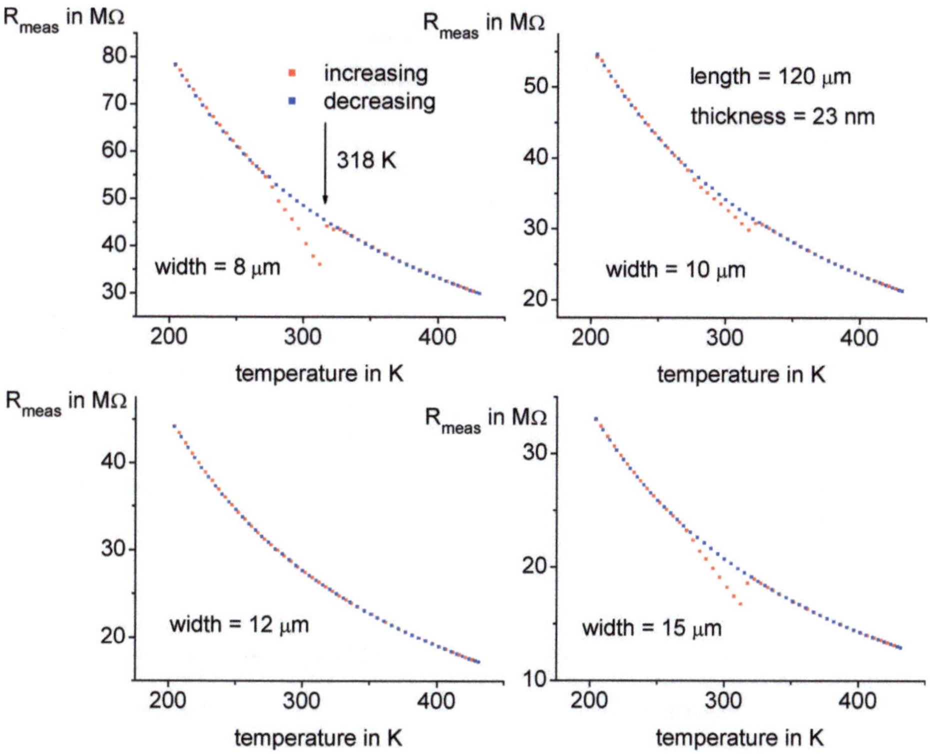

Figure 16.19: Thermal hysteresis and the chaos point.

Note: Illustration of the thermal hysteresis of a group of resistors of equal length (120 µm) but of different widths. The largest difference in behavior takes place at the chaos point at ≈45 °C (τ/3.57). Interestingly, a resistor of width of ≈12 µm shows no difference between increasing and decreasing measurement cycles. In other words, the thermal hysteresis effect in the resistivity is completely suppressed for ≈12 µm. Data provided with kind permission of Cicor Micro Electronics Reinhardt Microtech AG.

Interestingly, two resistors reveal no thermal hysteresis at approximately the same resistance width of ≈12 µm (Figure 16.19) or ≈13 µm (Figure 16.20). Does the absence of thermal hysteresis of both resistors have something to do with the characteristic length L (≈12.7 µm)?

16.4.3 Indications of the existence of a basic energy scale

As described in the previous chapters, the electrical resistivity measurement of amorphous materials as a function of temperature is subject to more or less pronounced dynamic nonequilibrium processes.

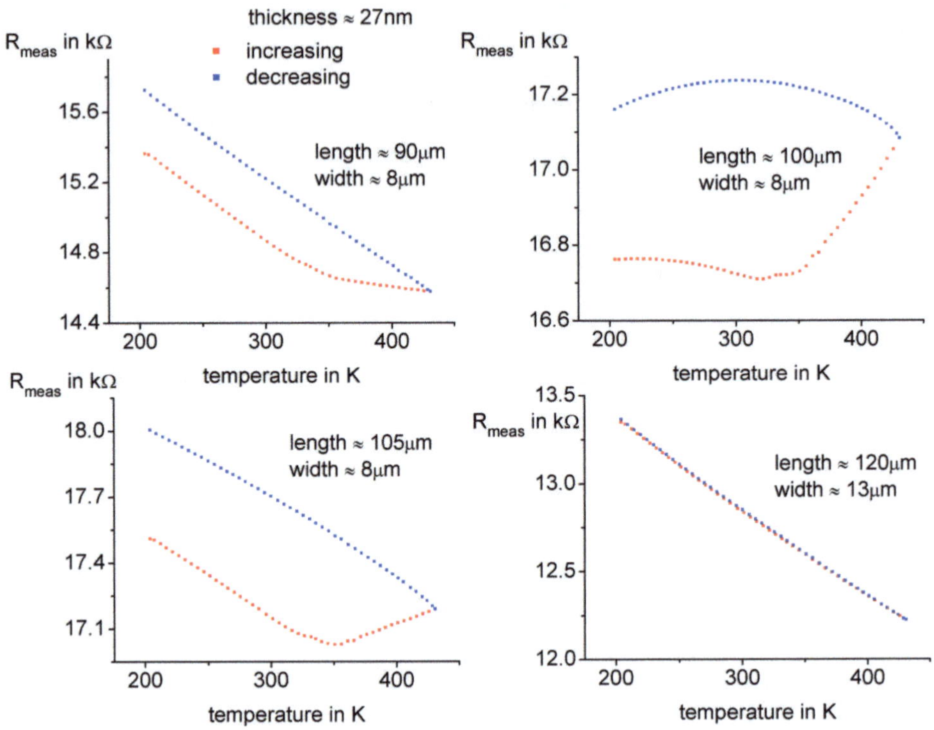

Figure 16.20: Thermal hysteresis and size effect.

Note: This representation shows a marked thermal hysteresis between heating and cooling cycles. In these cases as well, a considerable dependence on geometry can be observed. The initial curves are extremely unstable and change strongly during subsequent measuring cycles. Three curves for increasing measuring temperatures show a kink in the vicinity of 350 K, which could be related to the energy scale τ/π (≈361 K ≈85 °C). Data provided with kind permission of Cicor Micro Electronics Reinhardt Microtech AG.

What determines this behavior? Is a heat effect detectable even on measuring samples that do not show a pronounced thermal hysteresis? To clarify this question, many resistance measurements of amorphous material compositions have been examined for anisotropic effects as a function of temperature by normalizing resistivity values by means of a simple mathematical rule

Let the normalized difference $\Delta R(T)$ of a resistive value during a temperature cycle be defined by the following:

Definition 16.4 $\Delta R = (R_{down} - R_{up})/(R_{down} + R_{up})$

where $R_{up}(T)$ denotes the resistive value at increasing measuring temperature and $R_{down}(T)$ the value at decreasing measuring temperature.

With Definition 16.4, dimensionless values as a function of temperature are obtained as a quantitative measure of how charge carriers behave during a heating-cooling cycle. Looking for universal behavior at a specific temperature, it makes sense to add all the values calculated by Definition 16.4, regardless of the geometry or the material composition of the systems measured.

The sum ($\Sigma\Delta R$) of 1,500 normalized difference measurements in the temperature range from 20 °C to 200 °C is reproduced in Figure 16.21. These evaluations are based solely on measurements whose graphs of the heating and cooling process could not be distinguished with the naked eye. It is noticeable that the spectrum $\Delta R(T)$ has a plateau between ≈35 °C and ≈95 °C, and that the differences outside the plateau decrease markedly with increasing temperatures. At 85 °C, the sum spectrum shows a significant, interesting discontinuity.

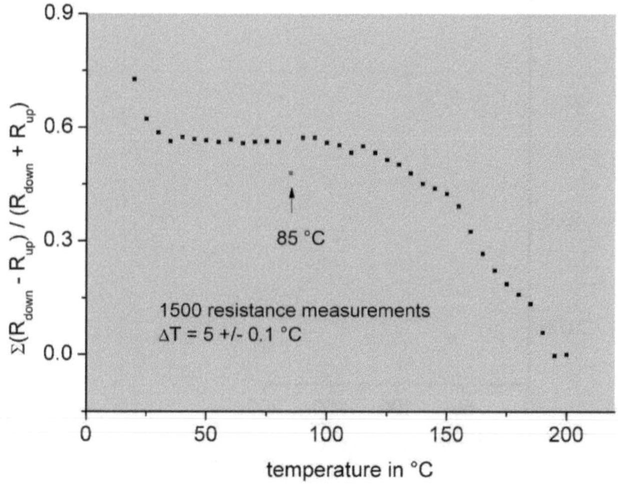

Figure 16.21: Resistive differences.

Note: Representation of the total amount of 1,500 normalized resistive differences $(R_{down}-R_{up})/(R_{down}+R_{up})$ of various samples as a function of temperature. The measuring temperature increment was ±5°C with an accuracy of ±0.1 °C. The irregularity observed at 85 °C (τ/π) is marked with an arrow. It clearly stands out against neighboring measuring points. Data provided with kind permission of Cicor Micro Electronics Reinhardt Microtech AG.

The temperature of 85 °C seems to be distinguished from all other measuring temperatures between 20 °C and 200 °C. If the transition temperature of 85 °C (358 K) is normalized by the reference temperature τ of 1,134 K, a dimensionless value of ≈ 3.168^{-1} is obtained, which corresponds approximately to π^{-1}.

What can be deduced from this information? Is the long-range magnetic energy scale $\tau_{bar} = \tau/(2\pi)$, introduced in Section 11.6, the seed of this irregularity? Or is it simply a measuring effect? Only further measurements with a higher temperature resolution can provide information on this matter. Interestingly, a resistive minimum of a measurement displayed in Figure 16.20 also occurred at 85 °C (358 K). Is there a connection with this minimum, or is it just a coincidence?

The "universal" discontinuity at ≈85 °C found by measuring various samples already occurred in one system when measuring the initial curve. At decreasing measuring temperatures, however, the abrupt transition at ≈85 °C with a resistance difference ΔR of ≈55,417 Ω was no longer visible in the curve shape and magically disappeared. It is interesting that the measured difference in the resistance ΔR is approximately equal to the sum $2(R_K+5\mu_0 c)$, or ≈ 55,393 Ω. The experimental situation is illustrated in Figure 16.22.

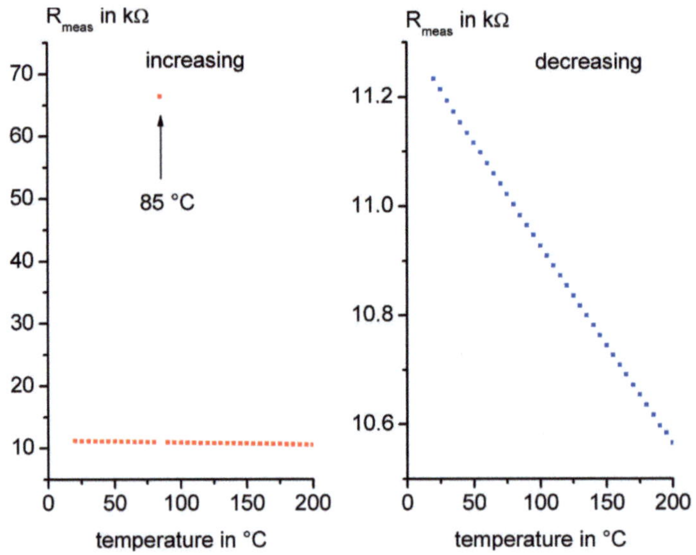

Figure 16.22: The discontinuity at 85 °C.

Note: The left diagram shows the raw data of an exceptional single measurement with a pronounced instability at 85 °C, which otherwise could only be observed by adding up many differential measurements. The right diagram exhibits the corresponding raw data of the cooling process. The measuring curve at decreasing temperatures follows almost exactly a linear characteristic within the measuring range with no discontinuity at 85 °C. Data provided with kind permission of Cicor Micro Electronics Reinhardt Microtech AG.

16.5 Emergent behavior of electrical transport

How can you predict the result when you can't predict what you will be measuring?
(*Philip Warren Anderson*)[7]

7 More and Different: notes from a thoughtful curmudgeon Philip W. Anderson, World Scientific Publishing Co. (2011)

The experiments listed in this section are actual events as measured and are neither glossed over nor constructed. In all cases, the macroscopically measured resistivity of the samples – similar to that shown in Figure 16.22 – was ambiguous, very strange and often against all reason. These observations are somewhat ghostly, because they were not reproducible, and responded very sensitive to the slightest disturbance. In a second measurement, the phenomena, which usually appeared as discontinuities[8] in the resistive measurements, mostly disappeared completely as if by magic. What is true, and what is untrue? A measurement error or an anomalous result with an experimental content regarding long-range interactions?

The measurement setup, that is, the interaction of the many-particle system, with the macroscopic measuring apparatus seems to play a role that should not be underestimated. Such vagaries of the experiment are difficult to operate for any experimenter because they are experimentally inaccessible and difficult to distinguish from artifacts. In short, a nightmare for any experimenter, since a reasonable analysis is very difficult. Such instabilities or emergences also do not permit statistical quantifications. They are either trivial measurement errors, for example, the contacting, or random, unstable physical processes that can sometimes be observed, but cannot, or only with difficulty, be influenced.

However, if these transport measurements, producing results that run counter to experience, are nevertheless any hints of nature? What do these "ghostly" measuring results tell us? Which conclusions can be drawn from it? Are they based on a collective, entangled behavior of many electrons involved, which eludes any control, is unexplainable and let alone calculable? Or is this unexpected behavior attributable to nonlinear waves, that is, to the spontaneous reaction of a system of charge carriers in a medium that is not in a thermodynamic equilibrium? Or are the electrons coupled to "background fields" that secretly dictate the long-range interactions of matter?

16.5.1 Measurements alpha

Figure 16.23 shows a resistance measurement where the initial curve changed abruptly at ≈137.5 °C, with a resistance difference ΔR of ≈31,762 Ω. In the entire cooling curve, no discontinuity is visible anymore. A second measurement cycle revealed no abnormal behavior in the resistance values for both heating and cooling. Particularly striking are the fluctuations in the resistance values of the initial curve between 20 °C and 35 °C, which differ drastically from those between 40 °C and 135 °C plotted in the small diagram. Of particular interest are the two peaks that are clearly visible at 25 °C

8 Discontinuities or phase transitions are a purely macroscopic property of matter. They do not occur in the microcosm of molecules.

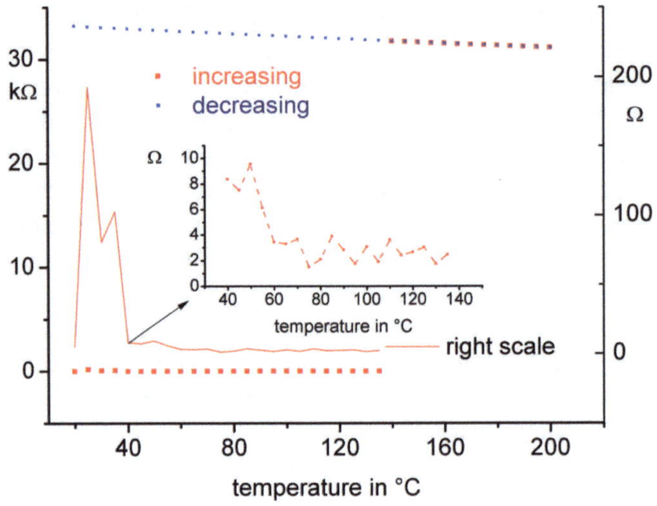

Figure 16.23: Measurement α1.

Note: Representation of the raw data of a measuring cycle between 20 °C and 200 °C. The resistance values, which were very low up to a temperature of 135 °C during the initial measurement cycle, increased abruptly at ≈137.5 °C and varied at decreasing measuring temperatures as usual according to a linear temperature characteristic. In the temperature range of low resistivity, the values measured at (30 ± 5) °C were significantly higher than elsewhere, indicating an additional instability in this range. The temperature range of very low resistivity is shown in a separate diagram. Data provided with kind permission of Cicor Micro Electronics Reinhardt Microtech AG.

and 35 °C. Is the peak at 35 °C possibly related to the Néel temperature of chromium of about 35 °C and thus also to the anomaly described in Section 16.4.1?
Of eight geometrically different resistors on the same substrate, only one resistor exhibited the aberrant behavior, as plotted in Figure 16.23. The measured thickness of the resistor is ≈21 nm, the width ≈40 µm [$\approx \pi L$ or $(\tau/\pi)^{-1}$ if $h = c = k_B = 1$] and the length ≈1,020 µm. This results in a specific resistance ρ_{el} of ≈2,620 µΩ cm at 140 °C and a specific resistance ρ_{el} of ≈0.2 µΩ cm at 135 °C, which is a factor of ten lower than the specific resistance of copper ρ_{Cu} of ≈ 1.7 µΩ cm at room temperature.

Using Ansatz 4.1, and setting $\lambda_F = a_B \approx 2.52$ nm and $(k_F l_e) = 1/4$, a specific resistance ρ_{el} according to the following relation

Relation 16.1 $\rho_{el} = R_K \lambda_F (k_F l_e)^{-1} \approx 258 (\lambda_F/\text{Å}) (k_F l_e)^{-1}$ µΩ cm ≈ 2,600 µΩ cm

can be calculated, which is close to the transition resistivity at 140 °C. If the measured transition difference ΔR of ≈31,762 Ω is divided by the von Klitzing constant R_K and the number constant π, that is, the ratio $\Delta R/R_K/\pi$ is formed, a value of ≈0.39 or ≈2/5 is obtained.

It may be coincidental that in the previous example the transition differential resistivity in units of R_K is a fractional multiple of π. However, such a coincidence was also observed in another measurement plotted in Figure 16.24. In this case, dividing

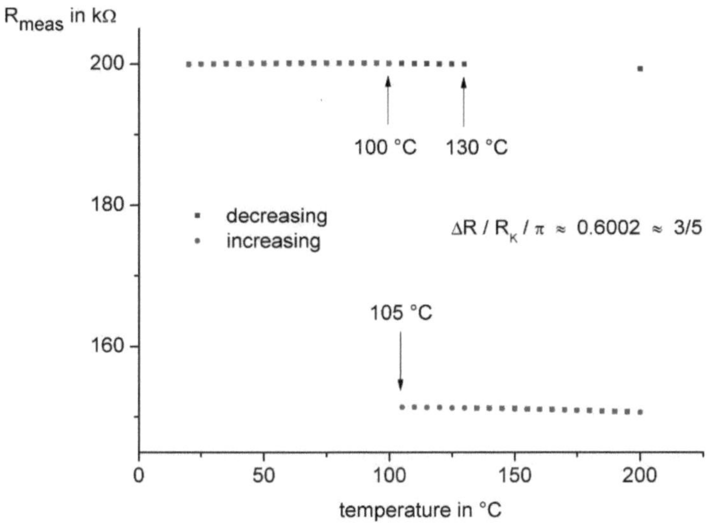

Figure 16.24: Measurement α2.

Note: Representation of abrupt transitions during a temperature measurement cycle on a hot plate. The observed change in resistance ΔR divided by the von Klitzing constant R_K and the mathematical constant π gives quite precisely the fraction 3/5. Interestingly, the discontinuity during the heating cycle occurred at the temperature τ/3 as given in Table 5.3. Data provided with kind permission of Cicor Micro Electronics Reinhardt Microtech AG.

the transition differential resistivity by the von Klitzing constant and the number π, the fraction 3/5 is attained with excellent accuracy.

16.5.2 Measurements beta

The two measurements α1 and α2 of Section 16.5.1 showed temperature curves with nearly linear temperature characteristics.

In Figure 16.25 a measurement is displayed where the measurement curves are strongly curved as a function of temperature. In this sample, the resistive transition occurred at a temperature of ≈112.5 °C. The initial curve shows metallic behavior from 20 °C to 110 °C with a positive temperature coefficient. At 110 °C, that is, prior to the transition, the resistivity is 3,200 μΩ cm, which equates to a sheet resistance of 1,454 Ω for a measured layer thickness of ≈22 nm. Immediately after the transition at ≈115 °C, the resistivity increased to ≈1.62 Ω cm, and the material behaved like a semiconductor with a negative temperature coefficient. Within 5 °C, the resistance thus changed by a factor of about 500. A second measurement revealed no abrupt resistive change anymore.

The ab-values of the initial curve (Section 16.2.1) between 20 °C and 110 °C are $+0.12 \cdot 10^{-3}\ K^{-1}$ (b) or $-95 \cdot 10^{-6}\ K^{-2}$ (a). For the cooling curve between 200 °C and 20 °C, they are $\approx -2.9 \cdot 10^{-3}\ K^{-1}$ (b) or $+2.6 \cdot 10^{-6}\ K^{-2}$ (a).

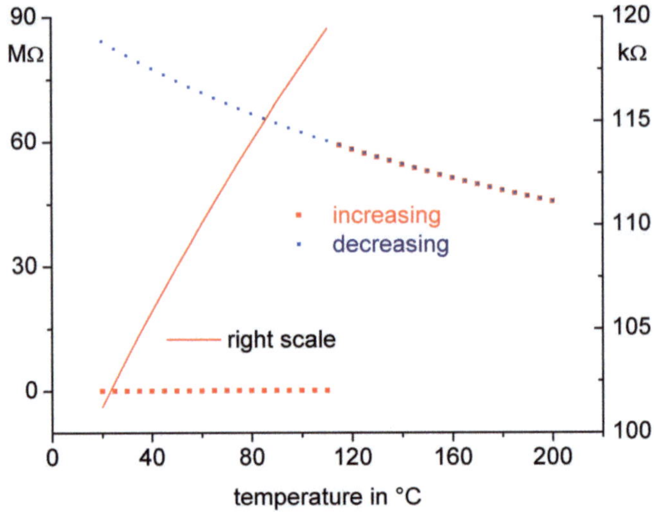

Figure 16.25: Measurement β1.

Note: An extreme change between metallic and nonmetallic behavior is visualized by this figure. At increasing measuring temperatures, the material behaves up to 110 °C similar to a metal. However, the temperature characteristic is not linear as usual for metals, but is slightly opened downward. After the transition, the resistance shows strongly localized behavior with a temperature characteristic opened upward, which is maintained throughout the cooling phase. Data provided with kind permission of Cicor Micro Electronics Reinhardt Microtech AG.

Using Ansatz 4.1, and setting $\lambda_F = a_B \approx 2.52\,\text{nm}$ and $(k_F l_e) = 1/5$, a specific resistance ρ_{el} according to

Relation 16.2 $\rho_{el} = R_K\,\lambda_F\,(k_F\,l_e)^{-1} \approx 258\,(\lambda_F/\text{Å})\,(k_F\,l_e)^{-1}\,\mu\Omega\,\text{cm} \approx 3{,}250\,\mu\Omega\,\text{cm}$

can be calculated. This value corresponds quite accurately to the maximum observed resistivity of 3,200 µΩ cm in the metal phase before the transition took place.

Figure 16.26 presents another measurement of a high-ohmic resistor, which changed the resistive value between ≈222.6 and ≈227.6 K from ≈44.3 to ≈327 MΩ. The transition could only be observed during the heating period and was no longer detectable in a second measurement. What caused this system to abruptly change to the more insulating phase after a short heating period?

16.5.3 Measurements gamma

In phase transitions, a physical quantity changes abruptly at a certain temperature. Such a behavior is not an unknown physical phenomenon. In most cases such changes are independent of how often the phase transition is crossed. However, for

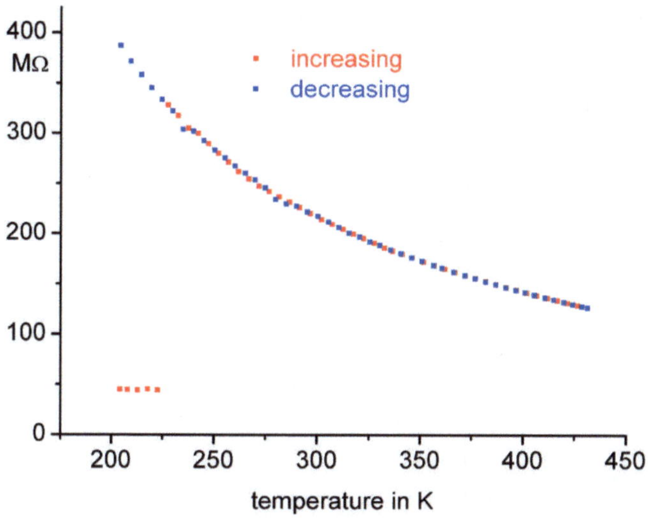

Figure 16.26: Measurement β2.

Note: This illustration shows the increase of the resistance value at ≈−50 °C by a factor of seven to eight. Similar to Figure 16.25, the temperature characteristic is comparable to the behavior of a semiconductor after the transition, that is, strongly curved and opened upward. Data provided with kind permission of Cicor Micro Electronics Reinhardt Microtech AG.

all the measurements described in Sections 16.5.1 and 16.5.2, the phase transitions occurred solely during the initial measurement, which does not correspond to a stable physical behavior. This is different with the example shown in Figure 16.27. Here, the discontinuity survived several measuring cycles and was also detectable on different instruments that were equipped with different ohmmeters. This gives some credibility to all observations, which were otherwise unique and spooky.

Figure 16.27(A) shows the initial curve of a resistor as a function of temperature with a metastable transition. The measurement was first performed with a probe card on a temperature-stabilized brass plate. From this it can be seen that at ≈130 °C the resistance abruptly changed by an amount of ≈856 Ω. The drastic change in resistance also happened almost exactly at the same temperature during a second measurement in a climatic chamber from −70 °C to 160 °C using a different measuring holder, that is, contacts, and another ohmmeter. The temperature profile of this measurement is shown in Figure 16.27(B). A further measurement with the same probe card on the brass plate gave the measuring curve as shown in Figure 16.27(C).

The resistance change of ≈856 Ω of the sample shown in Figure 16.27 is on the order of the characteristic impedance of the vacuum $\mu_0 c$ of ≈376.7 Ω. If the differential resistance of ≈856 Ω is divided by the free-space characteristic impedance $\mu_0 c$, a value of ≈2.27 or ≈9/4 is obtained. Interestingly, both the discontinuity of measurement γ1

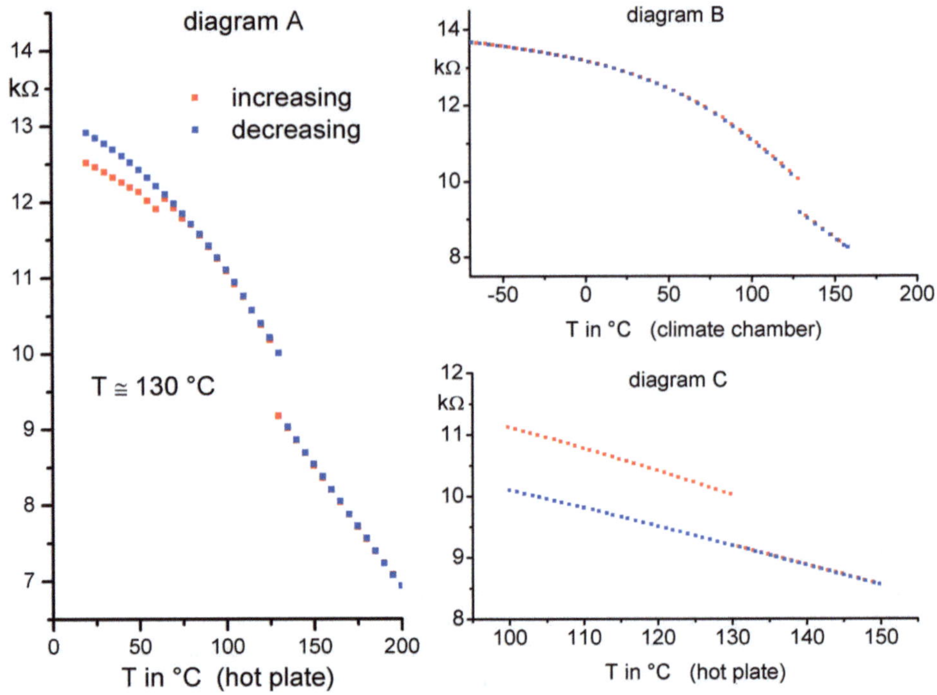

Figure 16.27: Measurement γ1.

Note: Measurements on a sample whose transport behavior could be reproduced by means of different measuring instruments: (A) shows the measured values of the first measuring cycle on a hot plate with a resistive discontinuity at ≈130 °C. The second measurement cycle, whose measured values are shown in (B), was carried out in a climate chamber from −70 °C to 160 °C. Apart from the discontinuity at the exact same temperature as in the first measurement cycle, the curve no longer shows any hysteresis. The third measurement cycle was performed again on the hot plate with a limited temperature range, as shown in (C). The resistance value at the same transition temperature changed by the same amount, but only with increasing measurement temperatures. The curve at decreasing measurement temperatures remains contiguous and reveals no discontinuity anymore. Data provided with kind permission of Cicor Micro Electronics Reinhardt Microtech AG.

and the discontinuity of measurement α2 occurred at the same transition temperature of ≈130 °C.

In the case of the quantum Hall effect, it is known that regardless of the material, the Hall resistance at very low temperature varies stepwise, that is, in fractions of the constant $R_K = \mu_0 c\,(2\alpha)^{-1}$, when the strength of the magnetic field is increased. The fact that discontinuities of the order of the von Klitzing constant R_K can occur not only at low but also at high temperatures is extremely unusual. Such a behavior is shown in Figure 16.28, which normally requires very low temperatures (near liquid helium temperature). After cooling to ≈−70 °C (starting point of the heating process)

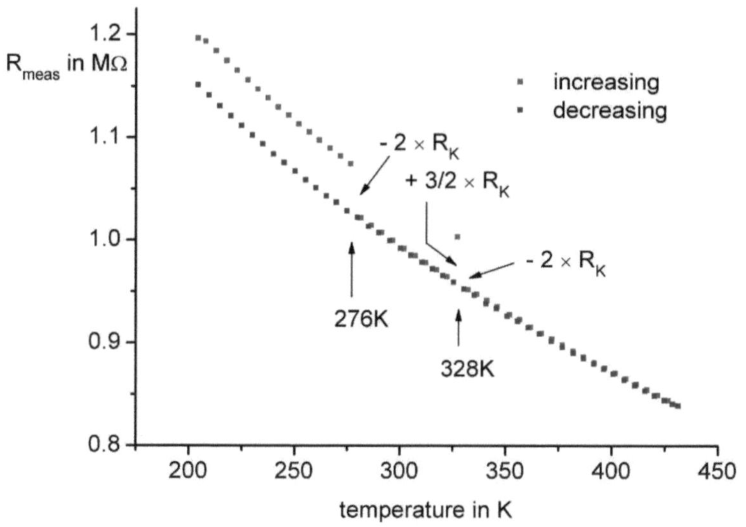

Figure 16.28: Measurement γ2.

Note: Discrete "fluctuations" of the initial curve on the order of R_K, which no longer occurred at decreasing measurement temperatures. A second measurement cycle, which is not shown in the figure, no longer showed any anomalous transitions, and all measured values corresponded to the values of the first measurement cycle for decreasing measurement temperatures. Data provided with kind permission of Cicor Micro Electronics Reinhardt Microtech AG.

this resistor abruptly changed its resistance value at ≈3 °C by ≈$-2R_K$ during heating, then briefly increased the resistance at ≈55 °C by ≈(3/2) R_K and lowered it again by ≈$-2R_K$ at ≈60 °C. Upon cooling, all discontinuities disappeared, and the resistor exhibited a normal, contiguous temperature profile, which was followed without discontinuities, when the resistor was cyclically measured again. It is unclear why the resistive changes in units of R_K are simple fractions and not fractions of π as observed in Section 16.5.1.

16.5.4 Measurements delta

Resistance is usually a material-specific quantity and effects that do not follow this rule are exotic exceptions. In the following, measurements are presented in which the resistive changes always were of the same value regardless of the material composition and the dimensions of the resistors measured.

In Figure 16.29, four typical examples of such resistance curves with discontinuities are illustrated; the resistive changes ΔR of which were always ≈45 kΩ, that is, approximately 7/4 in units of the von Klitzing constant R_K. The differential value 7/4 R_K was in all cases independent of the absolute resistance value of the resistor

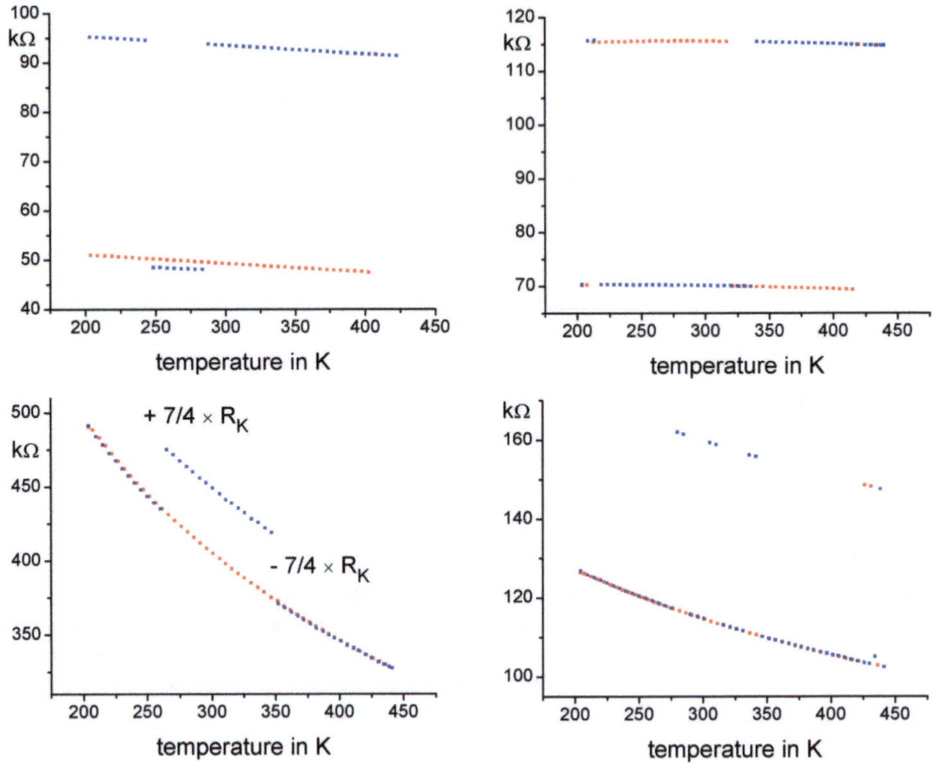

Figure 16.29: The discontinuity of 45 kΩ.

Note: Visualization of the raw data of four typical resistance measurements with resistance changes ΔR of ≈ 45 kΩ. By defining that the change in resistance ΔR corresponds to the difference of the measured value at higher temperature minus the measured value at lower temperature, a sign is assigned to the discontinuity. Part (C) takes this fact into account. Data provided with kind permission of Cicor Micro Electronics Reinhardt Microtech AG.

measured. In a second measurement, the discontinuities often disappeared, or they showed at least a very different temperature discontinuity pattern than in the initial measurement. The transitions did not take place at very specific temperatures, but seemed to extend randomly over the entire, measured temperature range. Only the evaluation of many measurements reflected that certain transition temperatures more frequently occurred than others.

If the resistor under investigation was replaced by a reference resistor, even the reference resistor was modified by ≈45 kΩ at certain temperatures. The anomalies suddenly disappeared without changing the measuring instrument. However, they also remained, if a new holder, new leads or freshly soldered connections were used. An obvious, simple explanation for the anomalies could not be found. Is the enigmatic phenomenon due to a misunderstood effect of the measurement setup? Why is the resistance difference, in almost all cases, about 7/4 in units of the von Klitzing constant R_K, regardless of the nature of the sample?

All detected discontinuities of many measured temperature profiles were stored in a database for statistical analysis. Data points were regarded as discontinuities, when the measurement curve as a function of temperature clearly revealed a discontinuous function that was visible to the naked eye. The resistance change ΔR was calculated as the difference of the measured value at higher temperature minus the measured value at lower temperature. This definition implies that each change in resistance is also assigned a sign. A positive ΔR can be interpreted as metallic behavior, since the resistance of typical metals increases with increasing temperature or decreases with decreasing temperature. Accordingly, a negative ΔR can be interpreted as nonmetallic behavior.

The normalization of a resistive change at a discontinuity is accomplished according to the following equation:

$$\Delta R_{norm} = \Delta R/R_K - 2\,\text{round}\,\{\Delta R/(2R_K)\}$$

This maps all resistance changes to the interval [–2, 2]. In Figure 16.30, the frequency distribution of all measured discontinuities is plotted with a subdivision of the interval [–2, 2] into 135 channels. The choice of the number of channels is arbitrary

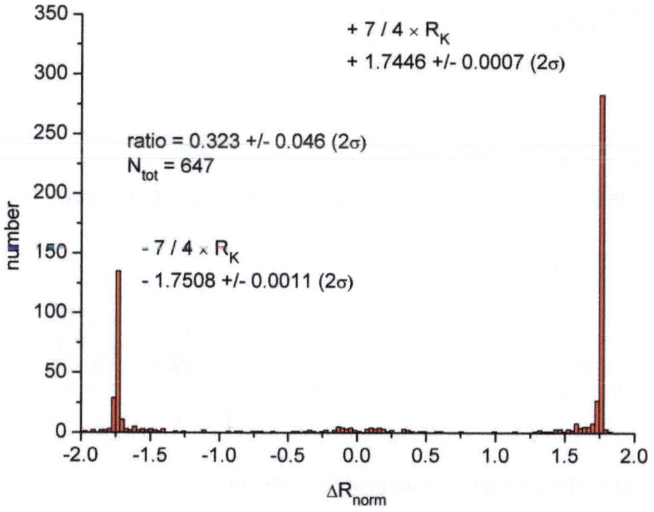

Figure 16.30: Frequency distribution of observed discontinuities.

Note: Representation of the frequency distribution of observed discontinuities with the normalized resistance change ΔR_{norm} as a characteristic. The number of channels is 135. About two-thirds of the discontinuities are concentrated at +7/4, and the other third at −7/4. Only a few discontinuities are distributed over the rest of the values in the interval [−2, 2]. Data provided with kind permission of Cicor Micro Electronics Reinhardt Microtech AG.

and consequently affects the histogram. By choosing the number of channels, the mean values indicated in the figure are also biased. Interestingly enough, the two largest frequencies are approximately in a ratio as 1:2. The major reason the number 135 was chosen is that 135 is a divisor of 1,890 (Section 3.3) and corresponds approximately to the reciprocal of the fine-structure constant.

Since the discontinuities occurred somewhere between two adjacent measurement temperatures, the transition temperatures T_c were calculated according to the formula $T_c = (T1+T2)/2$. In Figure 16.31, the frequency distribution of the transition temperatures T_c associated with Figure 16.30 is made visible with a temperature width of 5 °C. Figure 16.31(A) shows the temperature distribution of all the tabulated discontinuities, while Figure (B) and (C) only displays those with a resistance difference of +7/4 or −7/4 in units of the von Klitzing constant R_K. If both positive

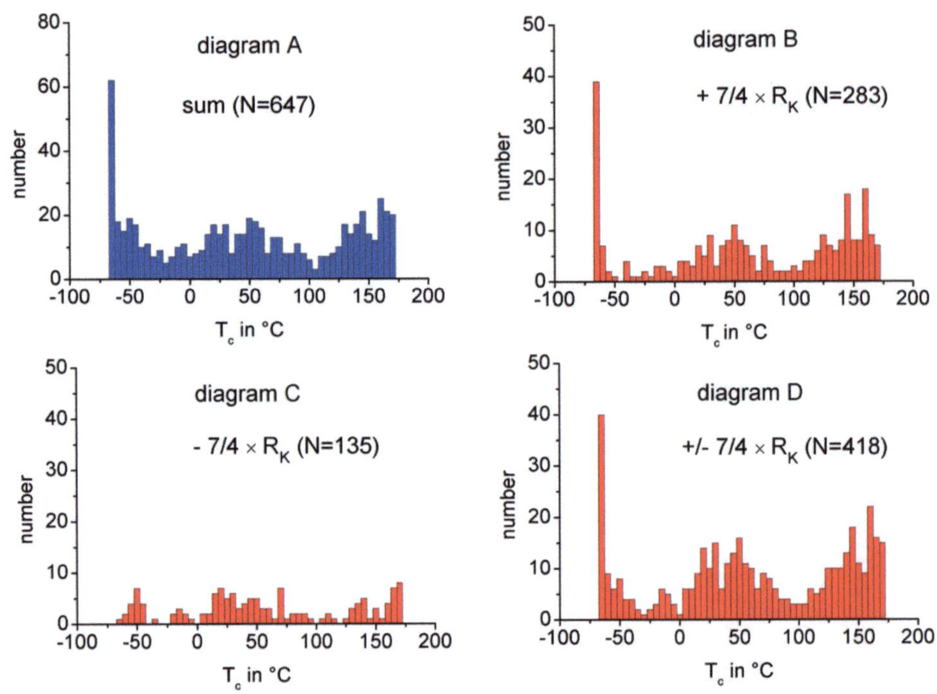

Figure 16.31: The T-dependence of the frequency distribution of Figure 16.30.

Note: Four histograms of the transition temperatures for resistors where resistive changes of +7/4, −7/4 and ±7/4 in units of the von Klitzing constant R_K took place. (B) This reflects that the discontinuities with a resistive change of +7/4 in units of the von Klitzing constant are not randomly distributed over the measured temperature range and occurred more frequently at the lowest temperature measured. While near the chaos point τ/3.57 (≈45 °C) strikingly more transitions occurred, in the vicinity of the bifurcation point τ/3 (≈105 °C) almost none exist. Data provided with kind permission of Cicor Micro Electronics Reinhardt Microtech AG.

and negative values (±7/4 RK) are taken into account, the frequency distribution of Figure 16.31(D) follows.

The observed discontinuities with a resistance difference of approximately +7/4 R_K (metallic behavior) are not completely random with regard to the temperature distribution. The transitions occur more frequently at certain temperatures, implying that the effect of metallic behavior must be temperature dependent. Striking is the large data accumulation at the smallest measurement temperature. By contrast, discontinuities with a resistance difference of –7/4 R_K (insulating behavior) appear to be more randomly distributed over the entire measurement range. Is the observed asymmetry a coincidence?

16.5.5 Measurement epsilon

In addition to the thermal hysteretic effect, a phenomenon appeared that is somehow linked to thermal hysteresis behavior.

After a measuring cycle most of the resistor samples were heat treated and the resistors measured again. For certain resistors, all of which exhibited a thermal hysteresis behavior analogous to the one shown in Figure 16.20, the resistance measurement after the heat treatment was sufficient to irreparably destroy the resistor. On the other hand, resistors that did not develop any thermal hysteresis on the same substrate, no damage could ever be observed. Since the thermal hysteresis strongly depends on the geometry of the resistor, it can be assumed that the damage of the resistors must be related to its dimensions as well.

Figure 16.32 shows an SEM image depicting a resistance damage. Two morphological changes in the resistive material can clearly be seen, which have formed channellike along the resistor edges. The ionization of the material along two parallel channels implies the existence of two current paths and thus the fact that the charge transport was not carried by the entire resistive material. The material has evidently been separated into nonconductive and conducting domains, and the current transport was limited to carriers that moved via self-organization within a restricted area. As a detail of Figure 16.32, the two damaged current channels are shown enlarged in Figure 16.33, in which the resistor width and the morphological changes in the material caused by the damage are also dimensioned.

The substrate on which the resistive material had been condensed is quartz. The measured thickness of the resistive layer is ≈30 nm. With a maximum possible measuring current of 10 mA, stipulated by the instrument used, a current density per current channel of ≈27 MA/cm^2 can be calculated with the aid of the data of Figure 16.33. Interestingly, this value is on the order of magnitude of the critical charge current density j_c of ≈49.6 MA/cm^2 from Section 5.7.1. Is that the reason why the resistor was damaged? Why were only resistors damaged that had an extremely strong thermal hysteresis? Is the experimentally observed hysteresis a consequence of the strong coupling of two conductive paths with different spin orientations?

Figure 16.32: The damage of a resistor.

Note: SEM image of the irreversible damage of a resistor after the measurement. Two channels parallel to the current direction are clearly visible, where the resistance material was destroyed by a measuring current that was too high. SEM image provided with kind permission of Cicor Micro Electronics Reinhardt Microtech AG.

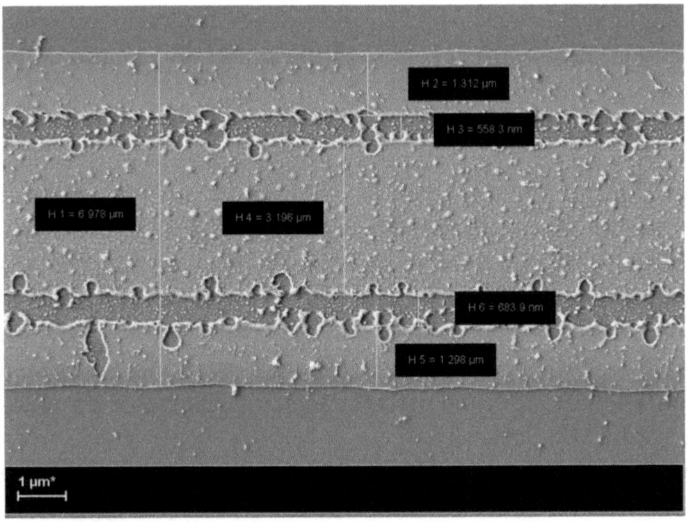

Figure 16.33: Enlarging detail of Figure 16.32.

Note: Enlarging detail of Figure 16.32 with dimensions of characteristic lengths. The average channel width of the two current paths, which are ≈ 3.2 µm apart, is approximately 600 nm. It is noteworthy that the distance between the two conductive channels is approximately equal to the length Λ. SEM image provided with kind permission of Cicor Micro Electronics Reinhardt Microtech AG.

16.5.6 Measurement zeta

If an electric current is impressed to a resistor, the response is usually a voltage drop in the direction of this current. The exact opposite response, namely, a voltage drop opposite to the direction of the applied current, gives a negative resistance, and is a very paradoxical effect that is usually considered as a measurement error. However, the measurement data in Figure 16.34 give rise to legitimate doubts that measurement ζ is an artifact, because a certain systematics can be recognized in the measurement data. Additionally, the polarity of the voltage can actually be measured very easily. But how can the directional movement be reversed at these temperatures? Why can the charge carriers suddenly move against the current direction?

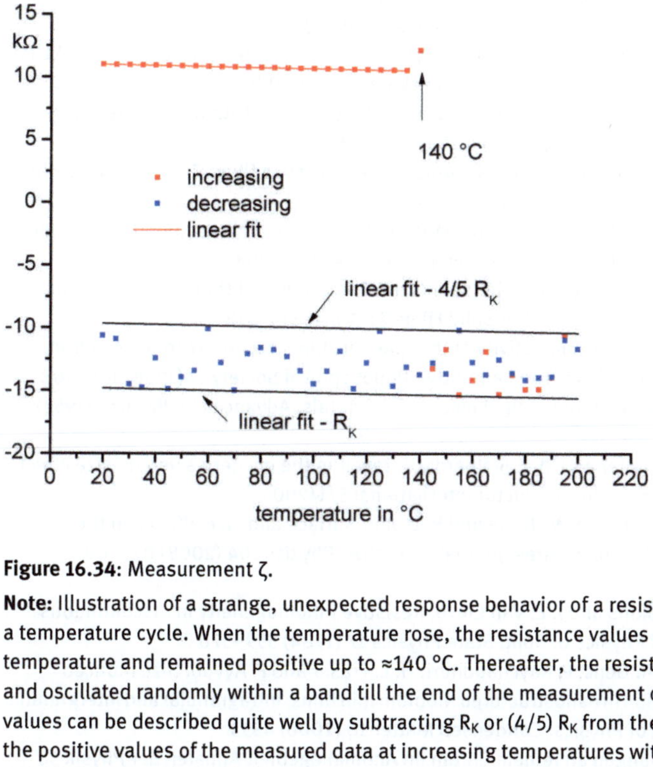

Figure 16.34: Measurement ζ.

Note: Illustration of a strange, unexpected response behavior of a resistance measurement during a temperature cycle. When the temperature rose, the resistance values decreased linearly with temperature and remained positive up to ≈140 °C. Thereafter, the resistance values became negative and oscillated randomly within a band till the end of the measurement cycle. The extreme negative values can be described quite well by subtracting R_K or (4/5) R_K from the best-fit straight line through the positive values of the measured data at increasing temperatures without using the data point at 140 °C. Data provided with kind permission of Cicor Micro Electronics Reinhardt Microtech AG.

The resistance values at increasing measuring temperatures behave "normally" between ≈20 °C and ≈135 °C and lie almost exactly on a straight line as a function of temperature. It is noteworthy that the fluctuations of the negative resistance values lie within a resistance band, which can be determined by subtracting (4/5) R_K or R_K from this straight line. When the same resistor was measured a second time, the effect was no longer measurable and the resistor behaved as it should be. All measured values of the second measurement lie on the straight line of the linear regression, which was obtained from the resistance values of the temperatures between 20 °C and 135 °C at increasing temperatures. The transition temperature at ≈140 °C is close to the zero-point energy (≈142 °C) of the mass m_0 within its Compton length $\Lambda = h/(cm_0)$. Is there possibly a connection between the two values?

References

[16.1] Charles Feldman. Temperature dependency of resistance of thin metal films. Journal of Applied Physics 34 (1963) 1710–1714

[16.2] A. A. Milgram and C. S. Lu. Electrical and structural properties of mixed chromium and silicon monoxide films. Journal of Applied Physics 39 (1968) 4219–4224

[16.3] G. J. Kahan. Equivalent circuit for conductivity temperature characteristics of the PdO/Ag-Pd glaze resistor. IBM Journal of Research and Development July 1971

[16.4] W. R. Hardy and D. K. Murti. Electrical and structural properties of NiCr thin film resistors reactively sputtered in O_2. Thin Solid Films 20 (1974) 345–362

[16.5] R. K. Waits. Silicon-chromium electron-beam-deposited resistive films. Transactions of the Metallurgical Society of AIME 242 (1968) 490–497

[16.6] A. J. Kushner and W. Worobey. Electrical resistance as a function of the temperature in pure and doped beta tantalum films. Thin Solid Films 23 (1974) 195– 203

[16.7] C. M. I. Birjega and S. Rau. Anomalies in the electrical resistance of thin NiCr films in the temperature range +25 to +80 °C. Thin Solid Films 21 (1974) S15–S18

[16.8] K. Fuchs and H. H. Wills. The conductivity of thin metallic films according to the electron theory of metals. Proceedings of the Cambridge Philosophical Society 34(1938) 100–108

[16.9] E. H. Sondheimer. The mean free path of electrons in metals. Advances in Physics 1(1952) 1–42

[16.10] W. Steinhoegl, G. Schindler and M. Engelhardt. Unraveling the mysteries behind size effects in metallization systems. Semiconductor International 5/1/2005

[16.11] Q. Huang, C. Lilley, M. Carmen, M. Bode and R. Divan. Surface and size effects on the electrical properties of Cu nanowires. Journal of Applied Physics 104 (2008) 023709 [6 pages]

[16.12] J. D. Benjamin, C. J. Adkins and J. E. Van Cleve. Resistive inhomogeneity in discontinuous metal films. Journal of Physics C: Solid State Physics 17 (1984) 559–576

[16.13] S. Bose, R. Banerjee, A. Genc, P. Raychaudhuri, H. L. Fraser and P. Ayyub. Size induced metal insulator transition in nanostructured Niobium thin films: Intragranular and intergranular contributions. Journal of Physics: Condensed Matter 18 (2006) 4553

[16.14] H. Söchtig. Untersuchungen an reinem Chrom im Anomaliegebiet. Annalen der Physik 38 (1940) 97–120

[16.15] S. Arajs, R. V. Colvin and M. J. Marcinkowski. Initial study of electrical resistivity of a chromium single crystal. Journal of the Less Common Metals 4 (1962) 46–51

[16.16] B. Stebler. The resistivity anomaly in chromium near the Neél temperature. Physica Scripta 2 (1970) 53–56

Appendix

A1 Special characters

a ~ b	a is proportional to b
a = b	a is equal to b
a ≈ b	a is approximately equal to b
a ≡ b	a is equal to b by definition

A2 Frequently used terms

Ansatz: The most obvious form of a mathematical relationship with variable parameters due to symmetries or physical relationships.

Condition: Assignment of values to variables for the determination of special states.

Definition: Introduction of a new variable or an auxiliary parameter.

Formula: Mathematical relation between physical quantities, which can be derived with the help of models. Proof of the relationship is not carried out. Often, however, reference is made to a citation where the context is explained in detail.

Equation: Equation of two terms for the determination of a variable.

Hypothesis: Relationship suspected on the basis of physical or mathematical interdependences, which carries the character of a law.

Relation: Relationship between physical quantities that can be derived from formulas, definitions or hypotheses.

A3 Physical units

Every physical quantity (observable) is the product of a numerical value with a corresponding unit. When written out, the following applies:

Physical quantity = numerical value × unit

In this book, all physical data are based on this syntax. The choice of another unit always amounts to shifting a numerical factor from the unit into the numerical coefficient, that is, the numerical value. The combination of a numerical value with a unit is an essential feature of all natural sciences, and the choice of the unit system is therefore of fundamental importance. Since measuring is always a comparing with reference values, this requires a simple, manageable system of units. Without referring

https://doi.org/10.1515/9783110612387-017

mathematical equations to quantitative values, no natural science is possible. The units used to measure physical quantities are however arbitrary. The choice of the units should not affect physical theories. All variables occurring in them are only scaled.

Throughout, the metric, decimal SI unit system has been deliberately used, which has proven itself in classical physics and above all in technology. This consistent system of units is based on seven basic units defined by convention in the sense of P. W. Bridgeman. Four of them establish the independent basic units of electrodynamics, which can be completely described by the meter m, the second s, the kilogram kg and the ampere A.

The three mechanical units s, m and kg are undoubtedly the three most important ones. The unit of charge is then the Coulomb C, that is, the ampere-second A s, and the unit of the electric field constant $\varepsilon_0 = 1/(\mu_0 c^2)$ is given by $kg^{-1}\,m^{-3}\,A^2\,s^4$. Both quantities are defined as fixed conversion factors by virtue of the fact that the magnetic field constant $\mu_0 \equiv 4\pi\,10^{-7}$ kg m $A^{-2}\,s^{-2}$ and the speed of light $c \equiv 299{,}792{,}458$ m s^{-1} are defined as fixed conversion factors. As a result, the proportionality factor of the Coulomb law is set equal to the dimensioned, fixed quantity $(4\pi\varepsilon_0)^{-1}$, similar to the consideration of Gauss, where the factor is 1. By setting the speed of light in vacuum equal to $c \equiv 299{,}792{,}458$ m s^{-1}, the meter is based on the time unit second. That is, the meter can be reproduced from the second by the relationship $m = 299{,}792{,}458^{-1}$ c s.

The Ampère is the only basic electric unit of the four basic units. Its definition comprises the equality of electrical and mechanical energies. By fixing the magnetic field constant μ_0, the ampere is actually unnecessary, as it allows the unit ampere to be ascribed to the mechanical units s, m and kg.

All the mechanical and electrical units known today are historically based on a nonrelativistic, classical physics with the following four values defined by SI convention: the prototype kilogram, the frequency of the cesium hyperfine transition $\Delta f(^{133}Cs)_{hfs}$ of exactly 9,192,631,770 s^{-1}, the speed of light c in vacuum and the magnetic field constant μ_0.

The ordinary use of "natural" units of particle physics is not being considered, with a small number of exceptions, although the relativistic equivalence between mass, energy and momentum is disguised. But it was precisely the strict compliance with the technical SI units that provided the "underlying theme" in the labyrinth of units and proved to be very expedient. The appropriate measurement system together with various models plays an essential role in this book. In some cases, the same symbols may be used for different physical quantities because they are deeply rooted in each discipline. However, the reader will easily notice from the context what quantity is meant.

The table of symbolic characters and their values summarizes all the most important quantities used in this book and lists newly introduced quantities using CODATA values [A 1]. CODATA values are not purely experimental data and are subjected to regular changes. They are model dependent (quantum electrodynamics) and strongly correlated with each other. The basic constants used in the calculations are marked in red; the quantities dependent thereof and calculated by means of a formula, in light red. The parameters required for the new view were calculated using CODATA values and are highlighted in yellow. All tabulated values are rounded to six decimal places.

A4 Symbolic characters and their values

Symbol	Meaning	Definition	Formula	Value	Unit
a_B	Relevant Bohr radius	Definition 2.2		2.515446 E−09	m
a_H	Hydrogen Bohr radius		$h/(2\pi\,\alpha\,c\,m_e)$	5.291772 E−11	m
c	Speed of light in vacuum	[A 1]	Exact	299,792,458	$m\,s^{-1}$
C	Heat capacity per volume				$J\,m^3\,K^{-1}$
c_1	Constant	Definition 3.1	$(4\,\pi^{4/3})^{-1/3}$	0.378755	
D	Density of states				$eV^{-1}\,m^{-3}$
E	Electric field strength				
E_B	Binding energy	Formula 2.1		0.286224	eV
E_{Fermi}	Fermi energy				
E_p	Plasma energy	Definition 7.3		0.068607	eV
E_T	Reference energy	Hypothesis 2.2		0.097715	eV
f	Frequency				
f_c	Cutoff frequency				
f_p	Plasma frequency	Definition 7.1		1.658921 E+13	s^{-1}
f_μ	Dimensionless factor	Definition 14.3			
G	Newton constant	[A 1]		6.67384 E−11	$m^3\,s^{-2}\,kg^{-1}$
g_e	Electron spin g-factor			2.002319304	
$G_F/(h_{bar}c)^3$	Fermi constant	[A 1]		1.166364 E−05	GeV^{-2}
h	Planck constant	[A 1]		6.626069 E−34	J s
h_{bar}	Reduced Planck constant		$h/(2\pi)$		
I	Current				
k_B	Boltzmann constant	[A 1]		1.380648 E−23	$J\,K^{-1}$
k_F	Fermi circular wave number				$rad\,m^{-1}$
k_{Ph}	Phonon circular wave number				$rad\,m^{-1}$
L	Universal length	Definition 5.2		1.267701 E−05	m

Symbol	Meaning	Definition	Formula	Value	Unit
L_1, L_2	Scattering lengths	Relation 8.1			
L_c	Cutoff length				
l_e	Electron coherence length				
L_{large}	Intergalactic length	Definition 6.3		5.515355 E+22	m
L_p	Plasma length	Definition 7.2		1.807153 E−05	m
l_{Ph}	Phonon coherence length				
L_{Planck}	Planck length	Definition 6.4	$h^{1/2}\, G^{1/2}\, (2\pi\, c^3)^{-1/2}$	1.616199 E−35	m
m, M	Masses				
m_0	Mass quantum	Hypothesis 2.1		5.102414 E−37	kg
m_e	Electron mass	[A 1]		9.109382 E−31	kg
m_{grav}	Fundamental mass scale	Equation 6.2		1.574137 E−27	kg
m_{Planck}	Planck mass				
m_W	$W^{+/-}$ boson mass	[A 2]		80.385 ± 0.015	GeV
m_Z	Z boson mass	[A 2]		91.1876 ± 0.0021	GeV
m_W/m_Z	Mass ratio	[A 2]		0.8819 ± 0.0012	
m_ν	Neutrino mass				
n	Particle number density	Formula 3.1		3.413723 E+18	cm^{-3}
$N_{Avogadro}$	Particle number per atomic mass unit		0.001 / u	6.022141 E+23	g^{-1}
p	Impulse				
P	Pressure				
Q	Total charge				
q_e	Elementary charge	[A 1]		1.602177 E−19	C
q_{Planck}	Planck charge	Formula 5.4			
r	Range, sphere radius				
R	Sphere radius				
r_e	Classical electron radius		$\alpha^2\, a_H$	2.817 940 E−15	m
R_{gas}	Universal gas constant				

r_{grav}	Gravitational length scale	Definition 6.7		2.234668 E−16	m
R_H	Hubble length				
rho	Relativistic mass density	Definition 6.1		4.427094 E−19	kg m^{-3}
R_K	von Klitzing constant		$h/q_e^2 = \mu_0 c/(2\alpha)$	25,812.807	Ω
R_{meas}	Measured resistance				
R_{sq}	Sheet resistance				
r_{rho}	Equilibrium radius	Definition 9.1		1.232439 E−07	m
R_y	Rydberg energy		$hcR_\infty = hc\,\alpha^2/(2\lambda_e)$		
T	Macroscopic temperature				
T_c	Critical temperature				
T_{Curie}	Ferromagnetic Curie temperature				
T_F	Fermi temperature of the free electron gas				
u	Atomic mass unit			1.660539 E−27	kg
U	Voltage				
u	Spectral energy density				J m^{-3} Hz^{-1}
V	Volume				
V_m	Volume of the m-dimensional hypersphere	Formula 3.6			
V_{sound}	Velocity of sound				
w	Total field energy				
w_{el}	Electrical field energy	Formula 9.4			
w_K	Electrical field temperature				
α	Sommerfeld constant	[A 1]	$q_e^2/(2\,h\,c\,\varepsilon_0)$	7.297352 E−03	
α_F	Dimensionless Fermi constant	Hypothesis 11.1		2.689311 E−10	
α_{grav}	Gravitational fine-structure constant	Definition 11.1		5.230749 E−39	
α_{geom}	Geometrized Sommerfeld constant	Definition 5.3	$\alpha_{geom} = 2^{-6}\,\pi^{-2/3}$	7.284282 E−03	
α_M	Magnetic coupling constant	Hypothesis 11.2		8.466410 E−19	
α_{mag}	Magnetic coupling constant	Definition 14.1			
α_{pol}	Polarizability				

Symbol	Meaning	Definition	Formula	Value	Unit
α_{WW}	Interaction coupling constant	Definition 14.4			
ε	Energy				
ε_0	Electric field constant	Exact	$10^7/(4\pi c^2)$	8.854188 E−12	$N^{-1} m^{-2} C^2$
ε_{mag}	Magnetic energy	Definition 14.2			
ε_r	Relative permittivity				
Θ_D	Debye temperature				
Θ_P	Paramagnetic Curie temperature				
κ	Specific thermal conductivity				
Λ	Compton length of m_0	Definition 2.1		4.331712 E−06	m
λ_0	Thermal wavelength	Formula 2.2			
λ_e	Compton length of m_e		$h/(m_e c) = 2\pi \alpha a_H$	2.426310 E−12	m
λ_{e_bar}	Reduced Compton length of m_e		$\lambda_e/(2\pi)$		
λ_F	Fermi wavelength				
λ_{Lo}	London length	Definition 5.1		2.098845 E−08	m
λ_{MI}	Metal–insulator length	Definition 4.1	$4\pi a_H$	6.649837 E−10	m
λ_{TF}	Thomas–Fermi shielding length	Formula 3.3			
μ_0	Magnetic field constant	[A 1]	Exact	$4\pi \cdot 1$ E−07	$N A^{-2}$
π	Circle constant			3.141593	
ρ	Mass density				
ρ_{el}	Specific electrical resistance				
σ	Stefan–Boltzmann constant		$2 \pi^5 k^4 h^{-3} c^{-2} 15^{-1}$	5.670373 E−08	$W m^{-2} K^{-4}$
τ	Reference temperature	Hypothesis 2.2		1,133.933	K
Φ	Coulomb potential				
χ	Magnetic susceptibility				
χ_e	Electrical susceptibility				
ω	Angular frequency		$\omega = 2\pi f$		$rad\ s^{-1}$

Basic constants are marked in red, and functions thereof in light red. The parameters with light yellow shading are introduced and explained in the references given. They all depend in some way on the basic constants.

A5 Page directory of approaches

A6 Page directory of definitions

A7 Page directory of formulas

A8 Page directory of relations

A9 Page directory of hypotheses, equations and conditions

References

[A 1] CODATA internationally recommended values of the fundamental. physical constants. http://physics.nist.gov/constants, queried on October 19, 2013

[A 2] J. Beringer et al. The review of particle physics. http://pdg.lbl.gov, queried on October 19, 2013

Register

https://doi.org/10.1515/9783110612387-018

Bei Fragen zur Produktsicherheit wenden Sie sich bitte an:
If you have any questions regarding product safety,
please contact:

Walter de Gruyter GmbH
Genthiner Straße 13
10785 Berlin
productsafety@degruyterbrill.com